Springer Proceedings in Mathematics & Statistics

Volume 513

This book series features volumes composed of selected contributions from workshops and conferences in all areas of current research in mathematics and statistics, including data science, operations research and optimization. In addition to an overall evaluation of the interest, scientific quality, and timeliness of each proposal at the hands of the publisher, individual contributions are all refereed to the high quality standards of leading journals in the field. Thus, this series provides the research community with well-edited, authoritative reports on developments in the most exciting areas of mathematical and statistical research today.

Juan Ramon Torregrosa
Editor

Mathematical Modelling in Engineering Human Behaviour

MME&HB 2024, Valencia, Spain, July 10–12

Springer

Editor
Juan Ramon Torregrosa
Instituto de Matemática Multidisciplinar
Universitat Politècnica de València
Valencia, Spain

ISSN 2194-1009 ISSN 2194-1017 (electronic)
Springer Proceedings in Mathematics & Statistics
ISBN 978-3-032-00566-3 ISBN 978-3-032-00567-0 (eBook)
https://doi.org/10.1007/978-3-032-00567-0

Mathematics Subject Classification: 82D75, 15B48, 34B45, 35R02, 37C75

This Springer imprint is published by the registered company Springer Nature Switzerland AG
The registered company address is: Gewerbestrasse 11, 6330 Cham, Switzerland

Preface

Mathematical modeling and numerical simulation play a key role in the scientific advances in science and engineering. The combination of mathematical language and the power of computers to carry out rapid and accurate computations is currently permitting a better understanding of complex phenomena that are of great interest to our society. The unifying approach to success in dealing with such challenges falls within the realm of mathematical modeling. Thanks to the mathematical methods that are continuously developing, researchers can gain insights into biological processes and medical diseases, predict outcomes in climate phenomena, and optimize designs in engineering mechanisms with unprecedented accuracy and efficiency.

This book collects some contributions dealing with new mathematical methods and applications covering a wide range of areas of Mathematics, providing an overview of hot topics where the mathematical community is currently advancing.

Indeed, the issue contains new results about continuous models formulated via differential equations and different methods for their numerical discretization. This area's contributions cover deterministic and stochastic techniques with applications to several related areas, including Biology, Ecology, Finance, Engineering, and so forth.

In this book, two chapters are devoted to matrix theory. In one of them, some properties of sign regular matrices, focusing on their connection to combined matrices, checkerboard matrices, and doubly quasi-stochastic matrices, are investigated. The main objective is to determine if there exists a nonsingular matrix in specific sign regular subclasses whose combined matrix is both doubly quasi-stochastic and strictly checkerboard or strictly inverse checkerboard. The other chapter about matrix theory establish a general procedure for determining the spectral characteristics, eigenvalues, and eigenvectors of Jacobi matrices with periodic coefficients. The technique is based on the analysis of difference equations with periodic coefficients and the associated boundary problems.

This book also contains two chapters dedicated to nuclear reactors and electrical networks. Classifying and localizing anomalies in nuclear reactors' performance is an issue to ensure the safety of nuclear reactors and, consequently, guarantee their long-term operations. In this work, the authors propose a deep learning framework for the deconvolution of reactor transfer functions from perturbation-induced neutron noise sources. The aim is to develop tools based on deep learning techniques using convolutional neural networks, working with simulated data with different Gaussian noise levels, and to study the number and location of detectors that need to be active.

The author shows that the problem of recovering the topology and admittance of an electrical network from power and voltage data at all vertices is often ill-posed, and sometimes it even has multiple solutions. So, the problem is reformulated to seek for a sparse network, i.e., with few edges, which fits the data up to a given tolerance. An algorithm for solving the problem is proposed, which combines the resolution of non-negative linear regression problems, and techniques of spectral graph sparsification.

On the other hand, a matter of mathematical biology is also presented. The liver has a great capacity to regenerate after partial hepatectomy but it is not always successful. Liver cells proliferate to replace the lost metabolic capacity. This work introduces two mathematical models based on ordinary differential equations for the regeneration of the liver after partial hepatectomy. The analysis of the model is limited by the complicated expressions for the steady solutions and their stability. Numerical simulations are done to study different scenarios.

The topic of combining the ordered weighted average operators and weighted average means has been extensively researched. This book also dedicates a chapter to explore how two independent weights of vectors can be merged to create a single operator that integrates characteristics from both weighted average means and, to a lesser extent, ordered weighted average. The resulting operator is expressed as a Choquet integral defined by a 2-additive capacity. In this way, the proposed index uses capacities without requiring further additional information.

Finally, the book present a new method for comparing rankings through complex networks: Model and analysis of competitiveness of the Spanish soccer league. As the author mentions, this method can be applied to other European soccer leagues or other sports.

We would like to thank Dr. Francesca Bonadei and the rest of the Editorial Staff for their generous disposition and sensitivity in supporting this publication.

Valencia, Spain Juan Ramon Torregrosa

Contents

On the Spectrum of Jacobi Matrices with Periodic Coefficients

A. M. Encinas, M. J. Jiménez, and S. Mondal

Abstract It is well known that for each strictly monotone list of n real numbers, there are infinitely many Jacobi matrices of order n whose eigenvalues are the entries of the list, along with a unique bisymmetric Jacobi matrix that satisfies this property. Families of such of matrices, with known spectra, are often used to test algorithms for eigenvalue determination or for recovering matrices from spectral data. Unfortunately, there are very few families of Jacobi matrices, even among bisymmetric ones, whose spectrum is known. In fact, for the specific class of Jacobi matrices with periodic coefficients, to our knowledge, the bisymmetric case has not been treated except for a few very particular examples. In this work, we seek to establish a general procedure for determining the spectral characteristics, eigenvalues, and eigenvectors of Jacobi matrices with periodic coefficients. We will then specialize the obtained formulas to the bisymmetric case for lower periods, up to 2. Our technique is based on the analysis of difference equations with periodic coefficients and the associated boundary problems.

Keywords Jacobi matrices · Bisymmetric matrices · Eigenvalues · Eigenvectors · Periodic coefficients · Sturm-Liouville difference equations

1 Introduction and Preliminaries

The inverse eigenvalue problem for symmetric matrices consists on determining whether for an ordered list of n real numbers there exists a symmetric and irreducible

A. M. Encinas (✉) · M. J. Jiménez
UPC, EEBE Barcelona, Barcelona, Spain
e-mail: andres.marcos.encinas@upc.edu

M. J. Jiménez
e-mail: maria.jose.jimenez@upc.edu

S. Mondal
University of Regina, Regina, Canada
e-mail: ma19d750@smail.iitm.ac.in; isamirmondal@gmail.com

J. R. Torregrosa (ed.), *Mathematical Modelling in Engineering Human Behaviour*,
Springer Proceedings in Mathematics & Statistics 513,
https://doi.org/10.1007/978-3-032-00567-0_1

matrix of order n whose spectrum coincides with the given list. Within this general problem, the case of bisymmetric Jacobi matrices occupies a central place, as for each strictly monotone list, there is a unique bisymmetric Jacobi matrix that realizes it, see [6, 17, 20, 21]. Here, the term *Jacobi matrix* refers to a symmetric tridiagonal matrix whose secondary diagonal has negative entries, making it an irreducible Z-matrix. The term *bisymmetric* signifies matrices that are symmetric with respect to their two main diagonals. Regardless of their significance in fields such as mechanics or statistics, families of this type of matrix with known spectra are frequently used to tests various recovery algorithms for determining the coefficients of the matrices from spectral data, see the comments in [16, Introducction]. Unfortunately, very few families of Jacobi matrices, including bisymmetric ones, have known spectra. Recently, researchers characterized bisymmetric Jacobi matrices whose eigenvalues form linear or quadratic monotone progressions, see [12]. This work brings together various studies conducted over the past hundred years. For the specific class of Jacobi matrices with periodic coefficients, results regarding the explicit computation of their eigenvalues primarily pertain to some cases of low periods, up to 2, see for instance [1, 24, 30], and moreover, to our knowledge, the bisymmetric case has not been treated.

In this work, we focus on obtaining general expressions for the spectrum of Jacobi matrices with periodic coefficients. Additionally, we explicitly describe the spectral characteristic for all bisymmetric Jacobi matrices of periodic coefficients with period up to 2. The obtained formulas serve as a specialization of a general procedure that first describes the eigenvalues and corresponding eigenvectors of any Jacobi matrix, and second, those of any Jacobi matrix with periodic coefficients. Our technique is based on the analysis of difference equations and the boundary (Sturm-Liouville) problems associated with them, as discussed in the general results of [10]. We also incorporate our findings regarding second-order difference equations with periodic coefficients (see [9]), providing a broader perspective on this class of matrices.

The main novelty of this work lies in establishing the study of the spectral characteristics of Jacobi matrices within the framework of boundary value problems for second-order difference equations. Specifically, we present here the following results:

1. Obtaining the spectral characteristic of Jacobi matrices from the fundamental solution of a Sturm-Liouville problem (Theorem 1).
2. Definition of the *almost-Toeplitz* case, a bit general than the usual Toeplitz one, and its analysis based on the Floquet theory for difference equations with periodic coefficients (Theorem 2). This explain the usual appearance of Chebyshev polynomials in the treatment of this kind of problems.
3. The introduction of the *compatible case*, that leads to relate eigenvalues with the zeroes of Chebyshev polynomials (Theorem 3), see also the Appendix.
4. The explicit derivation of the spectral characteristics for all bisymmetric Jacobi matrices with periodic coefficients of periods up to 2 (Sects. 4 and 5).

We emphasize that by applying the same techniques, it is possible to obtain the spectral characteristics of more general bisymmetric Jacobi matrices with periodic coefficients and in particular, those with periods up to 4. However, the richness and diversity of the cases to be studied have led us to restrict ourselves here to describing the general case and the explicit formulas for periods up to 2. The cases with periods 3 and 4 will be dealt with in later works.

Throughout the paper, vectors in $\mathbb{R}^{n+1}$, $n \in \mathbb{N}^*$, are denoted as $\mathsf{v} = (v_0, \dots, v_n)$ (always with the subindex starting at $k = 0$). With this notation, $\mathsf{v} > 0$ means that $v_k > 0$, $k = 0, \dots, n$ and the set of such vectors is denoted as $\mathbb{R}^{n+1}_+$. In addition, the null vector and the all ones vector in $\mathbb{R}^{n+1}$ are denote by $\mathsf{0}_{n+1}$ and e_{n+1}, respectively, although we drop the subindexes when it does not lead to confusion.

We call $\mathsf{z} \in \mathbb{R}^{n+1}$ *periodic with period* $p \in \mathbb{N}^*$, $1 \le p < n$ if it satisfies that $z_{p+k} = z_k$, for any $k = 0, \dots, n+1-p$. The subspace of periodic vectors in $\mathbb{R}^{n+1}$ with period p is denoted as $\mathbb{R}^{n+1}_p$. It is clear that $\dim(\mathbb{R}^{n+1}_p) = p$, since each periodic vector with period p is determined by its first p entries. Specifically, if $n + 1 = mp + r$ with $m \in \mathbb{N}^*$ and $0 \le r < p$, then $z_{kp+j} = z_j$ for any $k = 0, \dots, m-1$ and $j = 0, \dots, p-1$, and when $r \ge 1$, $z_{mp+j} = z_j$ for any $j = 0, \dots, r-1$. We also consider the set $(\mathbb{R}^{n+1}_+)_p = \mathbb{R}^{n+1}_+ \cap \mathbb{R}^{n+1}_p = \{\mathsf{z} \in \mathbb{R}^{n+1}_p : \mathsf{z} > 0\}$.

We call $\mathsf{z} \in \mathbb{R}^{n+2}$ *almost-periodic with period* $p \in \mathbb{N}^*$, $1 \le p < n$ if it satisfies that $z_{p+k} = z_k$, for any $k = 1, \dots, n-p$. The subspace of almost-periodic vectors in $\mathbb{R}^{n+2}$ with period p is denoted as $\mathbb{A}^{n+2}_p$ and then $\dim(\mathbb{A}^{n+2}_p) = p + 2$. Specifically, if $n + 1 = mp + r$ with $m \in \mathbb{N}^*$ and $0 \le r < p$, then $z_{kp+j} = z_j$ for any $k = 0, \dots, m-1$ and $j = 1, \dots, p-1$, and when $r \ge 1$, $z_{mp+j} = z_j$ for any $j = 0, \dots, r-1$. In addition, we define the *periodicity values* of a as $\sigma = z_0 - z_p$ and $\tau = z_{n+1} - z_r$. Of course, in any case, when $\mathsf{z} \in \mathbb{A}^{n+2}_p$, then $\mathsf{z} \in \mathbb{R}^{n+2}_p$ if and only if $\sigma = \tau = 0$.

Given $\ell \in \mathbb{N}^*$, we call $\mathsf{z} = (z_0, \dots, z_\ell) \in \mathbb{R}^{\ell+1}$ *centrosymmetric* when $z_{\ell-k} = z_k$, for all $k = 0, \dots, \ell$, or equivalently if $z_{\ell-k} = z_k$ for any $k = 0, \dots, \lfloor \frac{\ell}{2} \rfloor$. Therefore, if $\mathbb{S}^{\ell+1}$ is the subspace of centrosymmetric vectors in $\mathbb{R}^{\ell+1}$, then $\dim(\mathbb{S}^{\ell+1}) = \lceil \frac{\ell+1}{2} \rceil$.

Given $n \in \mathbb{N}^*$, the main objective of this work is to describe the eigenvalues and eigenvectors of some clases of Jacobi matrices of order $n + 2$. According with the above notation, Jacobi matrix means a tridiagonal matrix of the form

$$\mathsf{J}(\mathsf{a}, \mathsf{b}) = \begin{bmatrix} a_0 & -b_0 & 0 & \cdots & 0 \\ -b_0 & a_1 & -b_1 & \ddots & \vdots \\ 0 & \ddots & \ddots & \ddots & 0 \\ \vdots & \ddots & \ddots & a_n & -b_n \\ 0 & \cdots & 0 & -b_n & a_{n+1} \end{bmatrix} \tag{1}$$

where $\mathsf{a} = (a_0, a_1, \dots, a_n, a_{n+1}) \in \mathbb{R}^{n+2}$ and $\mathsf{b} = (b_0 \dots, b_n) \in \mathbb{R}^{n+1}_+$.

The Jacobi matrix $\mathsf{J}(\mathsf{a}, \mathsf{b})$ is called *bisymmetric* when a and b are both centrosymmetric; that is, $a_{n+1-k} = a_k$, $k = 0, \dots, n+1$ and $b_{n-k} = b_k$, $k = 0, \dots, n$. Therefore, a bisymmetric Jacobi matrix is determined by $\lceil \frac{n+1}{2} \rceil + \lceil \frac{n+2}{2} \rceil = n + 2$ values.

The Jacobi matrix $\mathsf{J}(\mathsf{a},\mathsf{b})$ is called *p -Toeplitz* when its entries along the diagonals are vectors of period p; that is, when $(\mathsf{a},\mathsf{b}) \in \mathbb{R}_p^{n+2} \times (\mathbb{R}_+^{n+1})_p$, see for instance [1, 24]. Attending this terminology, if $1 \le p < n$, we call the Jacobi matrix $\mathsf{J}(\mathsf{a},\mathsf{b})$ *almost p-Toeplitz* when a is almost-periodic with period p and b is periodic with period p; that is, when $(\mathsf{a},\mathsf{b}) \in \mathbb{A}_p^{n+2} \times (\mathbb{R}_+^{n+1})_p$. We call *Jacobi almost-Toeplitz* matrix a Jacobi matrix that is almost p-Toeplitz for some $p \ge 1$.

Remark 1 In [1], the Jacobi almost p-Toeplitz matrices are called *perturbations of Jacobi p -Toeplitz matrices*. In fact, in that paper, the authors choose to work with non-symmetric tridiagonal matrices whose upper secondary diagonal entries are positive and have low secondary diagonal entries equal to 1. Well known properties of irreducible tridiagonal matrices ensure that the eigenvalues are the same as those of a symmetric tridiagonal matrix.

Along the paper, we always assume that $0^0 = 1$ and also the usual conventions that empty products and empty sums are defined as 1 and 0 respectively.

2 Eigenvalues, Eigenvectors and Sturm-Liouville Problems

Given $\mathsf{f} \in \mathbb{R}^{n+2}$, $\mathsf{v} \in \mathbb{R}^{n+2}$ satisfies $\mathsf{J}(\mathsf{a},\mathsf{b})\mathsf{v} = \mathsf{f}$ iff it is a solution of the system

$$\begin{aligned} a_0 v_0 - b_0 v_1 &= f_0, \\ -b_{k-1} v_{k-1} + a_k v_k - b_k v_{k+1} &= f_k, \quad k = 1, \ldots, n, \\ -b_n v_n + a_{n+1} v_{n+1} &= f_{n+1}. \end{aligned} \tag{2}$$

In particular, when $\mathsf{f} = \mathsf{0}$ we refer to the above linear system (2) as *homogeneous*.

We are interested in describing those real numbers, $\lambda \in \mathbb{R}$, and those vectors, $\mathsf{v} \in \mathbb{R}^{n+2}$, such that $\mathsf{J}(\mathsf{a},\mathsf{b})\mathsf{v} = \lambda \mathsf{v}$ or equivalently, such that they are the solutions of the homogeneous system $\mathsf{J}(\mathsf{a} - \lambda \mathsf{e}, \mathsf{b})\mathsf{v} = \mathsf{0}$.

If for any $\lambda \in \mathbb{R}$ we consider the subspace $\mathcal{V}_{\mathsf{a},\mathsf{b}}(\lambda) = \ker\big(\mathsf{J}(\mathsf{a} - \lambda \mathsf{e}, \mathsf{b})\big)$, the *nullity of* $\mathsf{J}(\mathsf{a} - \lambda \mathsf{e}, \mathsf{b})$, λ is called *an eigenvalue of* $\mathsf{J}(\mathsf{a},\mathsf{b})$ when $\dim \mathcal{V}_{\mathsf{a},\mathsf{b}}(\lambda) \ge 1$ and then $\mathcal{V}_{\mathsf{a},\mathsf{b}}(\lambda)$ is called *eigenvector subspace* corresponding to λ. Therefore, the eigenvalues of $\mathsf{J}(\mathsf{a},\mathsf{b})$ are the roots of $P_{\mathsf{a},\mathsf{b}}(x) = \det\big(\mathsf{J}(\mathsf{a} - x\mathsf{e}, \mathsf{b})\big)$, the *characteristic polynomial* of $\mathsf{J}(\mathsf{a},\mathsf{b})$.

Remark 2 From the combinatorial point of view, the system (2) can be interpreted as the action of a *discrete Schrödinger operator with potential* q over a path on $n + 2$ *nodes* labelled as $0, 1, \ldots, n, n + 1$, being 0 and $n + 1$ the *boundary* nodes and $\{1, \ldots, n\}$ the *interior* nodes. The *edges* of the path are $e_k = \{k, k + 1\}, k = 0, \ldots, n$, and we consider that edge e_k carried the *weight* b_k (we can image that edge e_k joining nodes k and $k + 1$ represents a wire with conductance b_k). The *potential* $\mathsf{q} = (q_0, \ldots, q_{n+1}) \in \mathbb{R}^{n+2}$ is defined as $q_0 = a_0 - b_0$, $q_k = a_k - b_k - b_{k-1}$, $k = 1, \ldots, n$ and $q_{n+1} = a_{n+1} - b_n$. This combinatorial point of view has a long tradition and is that adopted in most of the works by some of the authors, see for instance [5].

As we can observe, in system (2), the equations for the boundary nodes are different from the equations corresponding to interior nodes of the path, and this is one of the main characteristics of our approach. In fact, we can view the equations at the interior nodes as a difference equation on the path, and the equations at the boundary nodes as two boundary conditions. Since each boundary condition only involves the boundary nodes and its adjacent nodes, they represent the discrete version of *separated*, also known as *Sturm-Liouville*, boundary conditions. This is the way we understand and study Jacobi matrices, interpreting them from (second order) discrete linear boundary value problems.

The linear operator $\Delta_{\mathsf{a},\mathsf{b}}\colon \mathbb{R}^{n+2} \longrightarrow \mathbb{R}^n$ defined by the three-term equation

$$\Delta_{\mathsf{a},\mathsf{b}}(\mathsf{v})_k = -b_{k-1}v_{k-1} + a_k v_k - b_k v_{k+1}, \quad k = 1, \ldots, n, \tag{3}$$

is named *discrete Schrödinger operator with coefficients* a *and* b. Given $\mathsf{f} \in \mathbb{R}^{n+2}$, the equation $\Delta_{\mathsf{a},\mathsf{b}}(\mathsf{v}) = \mathsf{f}$ on $\{1, \ldots, n\}$ is known as the *Schrödinger (difference) equation* with coefficients a and b and data f. Since in this work we only consider the homogeneous Schrödinger difference equations; that is, when $\mathsf{f} = \mathsf{0}$, in the sequel we use Schrödinger equation as synonymous of homogeneous Schrödinger equation.

Remark 3 Both, this finite version and its generalization to the infinite case; i.e. when $k \in \mathbb{N}$ or even $k \in \mathbb{Z}$, corresponding respectively to infinite or doubly infinite Jacobi matrices, have been profusely treated in the literature, see for instance the papers [3, 5, 7, 9, 11, 23, 27, 29] to mention just a few. When $b_k = 1$ for any $k \in \mathbb{Z}$, the homogeneous semiinfinite Schrödinger equation is named *Harper equation* and encompasses the so–called *almost Mathieu operator* that has deserved special attention not only by its connections to physics, but also for its rich spectral theory. In fact, one of the main problems in this area, related with the topological structure of the spectra and popularized as the *Ten Martini Problem*, was solved fifteen years ago by concatenating the work of many outstanding researchers, see for instance [2, 3, 28]. In particular, the case of the almost Mathieu operator with rational frequencies corresponds to the equation (3) with periodic coefficients.

The operator $\Delta_{\mathsf{a},\mathsf{b}}$ does not depend on the boundary entries, a_0 and a_{n+1}, of vector a. Now we clarify their role by considering the *boundary operators*, $\mathcal{B}^0_{\mathsf{a},\mathsf{b}}, \mathcal{B}^{n+1}_{\mathsf{a},\mathsf{b}}\colon \mathbb{R}^{n+2} \longrightarrow \mathbb{R}$ defined as

$$\mathcal{B}^0_{\mathsf{a},\mathsf{b}}(\mathsf{v}) = a_0 v_0 - b_0 v_1 \quad \text{and} \quad \mathcal{B}^{n+1}_{\mathsf{a},\mathsf{b}}(\mathsf{v}) = a_{n+1} v_{n+1} - b_n v_n \tag{4}$$

that depend only on the coefficients a_0, a_{n+1}, b_0 and b_n.

The homogeneous *Sturm-Liouville Problem* associated with a and b consist on finding those $\mathsf{v} \in \mathbb{R}^{n+2}$ satisfying

$$\Delta_{\mathsf{a},\mathsf{b}}(\mathsf{v}) = \mathsf{0} \quad \text{and} \quad \mathcal{B}^0_{\mathsf{a},\mathsf{b}}(\mathsf{v}) = \mathcal{B}^{n+1}_{\mathsf{a},\mathsf{b}}(\mathsf{v}) = 0. \tag{5}$$

If $\mathcal{V}_{\mathsf{a,b}} \subset \mathbb{R}^{n+2}$ is the set of solutions of (5), then it is clear that $\mathcal{V}_{\mathsf{a,b}}$ is a vector subspace of $\ker(\Delta_{\mathsf{a,b}})$, the nullity of $\Delta_{\mathsf{a,b}}$. The Sturm-Liouville problem is called *regular* when $\dim \mathcal{V}_{\mathsf{a,b}} = 0$ and *singular* otherwise.

The following Lemma summarizes the well-known properties of the solutions for the homogeneous Schrödinger equation; that is, the elements in $\ker(\Delta_{\mathsf{a,b}})$. All results and motivations can be found in [9, 10].

Lemma 1 *When the Sturm-Liouville problem* (5) *is singular, then* $\dim(\mathcal{V}_{\mathsf{a,b}}) = 1$ *and moreover,* $\mathcal{V}_{\mathsf{a,b}} = \mathrm{span}\{\mathsf{v}\}$, *where* v *is the unique solution of the homogeneous Schrodinger equation satisfying* $v_0 = b_0$, $v_1 = a_0$.

In view of the above result, given $(\mathsf{a}, \mathsf{b}) \in \mathbb{R}^{n+2} \times \mathbb{R}_+^{n+1}$, we call *fundamental solution* of the Schrödinger equation $\Delta_{\mathsf{a,b}}(\mathsf{z}) = 0$, the unique $\mathsf{v} \in \mathbb{R}^{n+2}$ such that $\Delta_{\mathsf{a,b}}(\mathsf{v}) = 0$ and moreover $v_0 = b_0$ and $v_1 = a_0$, which implies that $\mathcal{B}^0_{\mathsf{a,b}}(\mathsf{v}) = 0$. So, the boundary value problem (5) is singular iff the fundamental solution of the homogeneous Schrödinger equation satisfies that $0 = \mathcal{B}^{n+1}_{\mathsf{a,b}}(\mathsf{v}) = a_{n+1}v_{n+1} - b_n v_n$ and then, $\mathcal{V}_{\mathsf{a,b}} = \mathrm{span}\{\mathsf{v}\}$.

We are ready to establish the main result about the eigenvalues and eigenvectors for the Jacobi matrix $\mathsf{J}(\mathsf{a}, \mathsf{b})$.

Theorem 1 *Given* $\mathsf{a} \in \mathbb{R}^{n+2}$ *and* $\mathsf{b} \in \mathbb{R}_+^{n+1}$, *for any* $x \in \mathbb{R}$ *consider* $\mathsf{v}(x)$ *the fundamental solution for the homogeneus Schrödinder equation* $\Delta_{\mathsf{a}-x\mathsf{e},\mathsf{b}}(\mathsf{z}) = 0$. *Then, for any* $k = 0, \dots, n+1$, $v_k(x)$ *is a polynomial in* x *with degree* k *and the characteristic polinomial of* $\mathsf{J}(\mathsf{a}, \mathsf{b})$ *is*

$$P_{\mathsf{a,b}}(x) = b_1 \cdots b_n\big((a_{n+1} - x)v_{n+1}(x) - b_n v_n(x)\big).$$

Therefore, $\lambda \in \mathbb{R}$ *is an eigenvalue of* $\mathsf{J}(\mathsf{a}, \mathsf{b})$ *iff* $(a_{n+1} - \lambda)v_{n+1}(\lambda) = b_n v_n(\lambda)$ *and when this happens* $\mathcal{V}_{\mathsf{a,b}}(\lambda) = \mathrm{span}\{\mathsf{v}(\lambda)\}$.

Proof We first prove that for any $k = 0, \dots, n+1$, $v_k(x)$ is a polynomial in x with degree k and principal coeficient $(-1)^k \prod_{j=1}^{k-1} b_j^{-1}$. It is true for $k = 0, 1$, since $v_0(x) = b_0$ and $v_1(x) = a_0 - x$ (remember that empty products equals 1). Assuming the claim is true for $1 \le k < n+1$, then $b_k v_{k+1}(x) = (a_k - x)v_k(x) - b_{k-1}v_{k-1}(x)$ and hence $v_{k+1}(x)$ is a polynomial with degree $k+1$ whose principal coeficient is $-b_k^{-1}$ times the principal coefficient of $v_k(x)$; that is $-b_k^{-1}(-1)^k \prod_{j=1}^{k-1} b_j^{-1} = (-1)^{k+1} \prod_{j=1}^{k} b_j^{-1}$. In conclusion, $b_1 \cdots b_n\big((a_{n+1} - x)v_{n+1}(x) - b_n v_n(x)\big)$ is a polynomial with degree $n+2$ and principal coefficient $(-1)^{n+2}$.

As λ is an eigenvalue of $\mathsf{J}(\mathsf{a}, \mathsf{b})$ iff $(a_{n+1} - \lambda)v_{n+1}(\lambda) - b_n v_n(\lambda) = 0$, we conclude that the above coincides with the characteristic polynomial of $\mathsf{J}(\mathsf{a}, \mathsf{b})$. □

Remark 4 If for any $k = 1, \dots, n+1$ we denote by $\mathsf{J}_k(\mathsf{a}, \mathsf{b})$ the Jacobi matrix obtained from $\mathsf{J}(\mathsf{a}, \mathsf{b})$ deleting the last $n+1-k$ rows and columns, it is easy to conclude that $P_k(x) = b_1 \cdots b_{k-1}\big((a_k - x)v_k(x) - b_{k-1}v_{k-1}(x)\big)$ is the characteristic polynomial of $\mathsf{J}_k(\mathsf{a},\mathsf{b})$ and that $P_{n+1}(x) = P_{\mathsf{a},\mathsf{b}}(x)$. Moreover, since $\det\big(\mathsf{J}_k(\mathsf{a},\mathsf{b})\big) = P_k(0)$, the result in Theorem 1 also leads us to identity

$$\det\big(\mathsf{J}_k(\mathsf{a},\mathsf{b})\big) = b_1 \cdots b_{k-1}\big(a_k v_k - b_{k-1} v_{k-1}\big), \tag{6}$$

where $\mathsf{v} \in \mathbb{R}^{n+2}$ is the fundamental solution of the equation $\Delta_{\mathsf{a},\mathsf{b}}(\mathsf{z}) = \mathsf{0}$.

After Theorem 1, it is clear that obtaining a closed formula for the eigenvalues of a Jacobi matrix is equivalent to give an expression for the fundamental solution of second order difference equations. So, the following question appears in a natural way:

For which coefficients $\mathsf{a} = (a_0, \dots, a_{n+1}) \in \mathbb{R}^{n+2}$ and $\mathsf{b} = (b_0, \dots, b_n) \in \mathbb{R}_+^{n+1}$ we can obtain explicitly $\mathsf{v}(x)$, the fundamental solution of the homogeneous Schrödinger equation

$$\Delta_{\mathsf{a}-x\mathsf{e},\mathsf{b}}(\mathsf{z})_k = -b_{k-1}z_{k-1} + (a_k - x)z_k - b_k z_{k+1} = 0, \quad k = 1, \dots, n, \tag{7}$$

for any $x \in \mathbb{R}$?

Recall that $\mathsf{v}(x)$ is the unique solution of Eq. (7) satisfying that $v_0(x) = b_0$ and $v_1(x) = a_0 - x$. Furthermore, the entry a_{n+1} of a has no role to compute $\mathsf{v}(x)$. This entry is relevant only when we need to know if x is an eigenvalue of $\mathsf{J}(\mathsf{a},\mathsf{b})$.

3 Spectral Characteristics of Jacobi Almost-Toeplitz Matrices

To treat to answer the question raised at the end of the previous section, we start our findings with the easier, and very well-known, case when the coefficients of the homogeneous Schödinger equation (7) are constant. This means that $a_k = a$, $k = 1, \dots, n$, and $b_k = b > 0$, $k = 0, \dots, n$.

The analysis of the constant coefficients case will serve us a motivation and guide for the following developments. Although a very common method is to consider the *Binet basis*; that is, the two roots of the polynomial $-bx^2 + (a - \lambda)x - b$, we follow here another route that was also used, implicitly or explicitly, in previous works, see for instance [1, 9] to mention only a few.

If for given $x \in \mathbb{R}$ we define $q = \dfrac{a - x}{2b}$, the homogeneous Schrödinger equation (7) is equivalent, in the sense that it has the same solutions, to the *Chebyshev equation with coefficient q*

$$-z_{k-1} + 2q z_k - z_{k+1} = 0, \quad k = 1, \dots, n. \tag{8}$$

Observe that if $\{P_k(x)\}_{k\in\mathbb{Z}}$ is a Chebyshev sequence of polynomials, see the Appendix, then $\mathsf{v} = \big(P_0(q), \dots, P_{n+1}(q)\big) \in \mathbb{R}^{n+2}$ is a solution of the Eq. (8).

As we see below, when the coefficients are non constant but they are periodic, then we can obtain the solutions of (7) in a similar way than in the case of constant coefficients. The authors are firmly convinced that this is the (technical) reason why so many papers on Jacobi matrices with periodic coefficients have appeared over time and also the reason for that Chebyshev polynomials play a fundamental role in the spectral analysis of this type of matrices. Recall that given $k \in \mathbb{Z}$, the *second kind Chebyshev polynomials*, are defined as $\{U_k(z)\}_{k\in\mathbb{Z}}$ the unique Chebyshev sequence determined from the three-term recurrence $-U_{k-1}(z) + 2zU_k(z) - U_{k+1}(z) = 0$ and satisfying that $U_0(z) = 1$ and $U_1(z) = 2z$. The main properties of this kind of polynomials are described in the Appendix, see Sect. 6.

Next we reproduce here the main results about linear second order difference equations with periodic coefficients with the aim of obtaining the fundamental solution of the corresponding homogeneous Schrödinger equation (7). All results related with difference equations with periodic coefficients can be found in [9].

We say that the Schrödinger operator $\Delta_{\mathsf{a},\mathsf{b}}$ *has periodic coefficients with period* p if $\mathsf{a} \in \mathbb{A}_p^{n+2}$ and $\mathsf{b} \in (\mathbb{R}_+^{n+1})_p$. It is clear that in this case, for any $x \in \mathbb{R}$ we have that $\Delta_{\mathsf{a}-x\mathsf{e},\mathsf{b}}$ has also periodic coefficients with period p.

In the sequel in this section, we consider fixed $n \in \mathbb{N}^*$, $1 \le p < n$ and $m = \lfloor \frac{n+1}{p} \rfloor$, so $m \in \mathbb{N}^*$ and $n + 1 = mp + r$, where and $0 \le r < p$; that is, $n + 1 = r(\bmod p)$. Notice that when $p = 1$, then $r = 0$, $m = n + 1$ and the case corresponds to a Schrödinger equation with constant coefficients, previously analized.

Although this scenario seems to be far away from the easiest one, we will show that it is not the case, since the main result in [9] establishes that irreducible Schrödinger equations (not necessarily self-adjoint as those here considered) with periodic coefficients are basically equivalent to some Chebyshev equation.

To understand the structure of the solutions of a second order linear, irreducible and self-adjoint difference equation with periodic coefficients of period p, it is useful to split a solution sequence $v_0, \ldots, v_{n+1}$ into the following p subsequences:

$$\begin{aligned} &v_{kp+j},\ k = 0, \ldots, m, &&\text{if } j = 0, \ldots, r, \\ &v_{kp+j},\ k = 0, \ldots, m-1, &&\text{if } j = r+1, \ldots, p-1. \end{aligned} \tag{9}$$

Therefore, the sequence $v_0, \ldots, v_{n+1}$ is splitted into $r + 1$ subsequences with $m + 1$ entries and $p - 1 - r$ subsequences with m entries. Of course, when $r = p - 1$, all subsequeces have $m + 1$ entries.

The structure of the solutions for difference equations with periodic coefficients is contained in the following result:

Lemma 2 ([9, Theorem 3.3]) *Let* $n \in \mathbb{N}^*$ *and* $1 \le p < n$, *and consider* $m \in \mathbb{N}^*$ *and* $0 \le r < p$ *such that* $n + 1 = mp + r$. *Then, there exists* $q_p \colon \mathbb{A}_p^{n+2} \times (\mathbb{R}_+^{n+1})_p \longrightarrow \mathbb{R}$ *satisfying that for any* $\mathsf{a} \in \mathbb{A}_p^{n+2}$ *and* $\mathsf{b} \in (\mathbb{R}_+^{n+1})_p$, $\mathsf{z} \in \mathbb{R}^{n+2}$ *is a solution of the homogeneous Schrödinger equation*

$$-b_{k-1}z_{k-1} + a_k z_k - b_k z_{k+1} = 0, \quad k = 1, \ldots, n,$$

iff for any $j = 0, \ldots, r$, z_{kp+j} *is a solution of the Chebyshev equation*

$$-z_{k-1} + 2q_p(\mathsf{a}, \mathsf{b})z_k - z_{k+1} = 0, \quad k = 1, \ldots, m,$$

and for any $j = r+1, \ldots, p-1$, z_{kp+j} *is a solution of the Chebyshev equation*

$$-z_{k-1} + 2q_p(\mathsf{a}, \mathsf{b})z_k - z_{k+1} = 0, \quad k = 1, \ldots, m-1.$$

Therefore, if $\mathsf{z} \in \mathbb{R}^{n+2}$ *satisfies that* $\Delta_{\mathsf{a},\mathsf{b}}(\mathsf{z}) = 0$*, then*

$$z_{kp+j} = z_{p+j}U_{k-1}\big(q_p(\mathsf{a}, \mathsf{b})\big) - z_j U_{k-2}\big(q_p(\mathsf{a}, \mathsf{b})\big)$$

where $k = 0, \ldots, m$, *if* $j = 0, \ldots, r$, *and* $k = 0, \ldots, m-1$, *if* $j = r+1, \ldots, p-1$.

While we know that any solution of the linear difference equation (7) is *implicitly* determined by its two first values, z_0 and z_1 by using the three-terms recurrence, the above result says that the knowledge of the first $2p$ values of any solution of a homogeneous Schrödinger equation with periodic coefficients of period p *explicitly* determines its remainder values. Of course, and as we have previously showed, this result is well-known when $p = 1$; that is, for constant coefficients. However, when $p \geq 2$, Lemma 2 showed that once this $2p$ values are determined, we get the explicit expression for the solutions. This $2p$ values can be obtained from the two first one; z_0 and z_1 by applying the three-term recurrence

$$\begin{aligned} z_k &= b_{k-1}^{-1}\big(a_{k-1}z_{k-1} - b_{k-2}z_{k-2}\big), && k = 2, \ldots, p, && \text{when } p \geq 2, \\ z_{p+1} &= b_0^{-1}\big(a_p z_p - b_{p-1}z_{p-1}\big), \\ z_{p+k} &= b_{k-1}^{-1}\big(a_{k-1}z_{p+k-1} - b_{k-2}z_{p+k-2}\big), && k = 2, \ldots, p-1, && \text{when } p \geq 3, \end{aligned}$$

where we have applied the periodicity of $a's$ and $b's$. Of course, this result will be efficient when $p \ll n$ and for the cases in which the explicit computation of the constant $q_p(\mathsf{a}, \mathsf{b})$ is also factible, usually for small values of p.

The function q_p is named *Floquet function* and its value $q_p(\mathsf{a}, \mathsf{b})$ at $\mathsf{a} \in \mathbb{A}_p^{n+2}$ and $\mathsf{b} \in (\mathbb{R}^+)_p^{n+1}$ is called *Floquet constant for the vectors* a *and* b. Therefore, obtaining the fundamental solution for a second order linear difference equation with periodic coefficients depends only on the knowledge of the associated Floquet constant. Unfortunately, since in [9, Theorem 3.3] the existence of this parameter was proven by induction, at first the above result does not seem to be useful in practice. For this reason, most of the aforementioned paper was devoted to the explicit computation of the *Floquet function*. Next we summarize all the steps to determine the Floquet function showing its combinatorial nature, see [9] for the complete details and also [11] in which this short summary is strongly based.

A *binary multi-index of order* p is a p-tuple $\alpha = (\alpha_0, \dots, \alpha_{p-1}) \in \{0, 1\}^p$ and its *length* is defined as $|\alpha| = \sum\limits_{j=0}^{p-1} \alpha_j \le p$. So $|\alpha| = m$ iff exactly m components of α are equal to 1 and exactly $p - m$ components of α are equal to 0.

Given $p \in \mathbb{N}^*$, we define the following subsets of $\{0, 1\}^p$:

(i) $\Lambda_p^0 = \big\{(0, \dots, 0)\big\}$, for $p \ge 1$.
(ii) $\Lambda_p^1 = \big\{\alpha : |\alpha| = 1\big\}$, for $p \ge 2$.
(iii) $\Lambda_p^m = \big\{\alpha : |\alpha| = m, \quad i_{j+1} - i_j \ge 2, \quad 1 \le j \le m - 1$ and $i_m \le p - 2$ if $i_1 = 0\big\}$ for $p \ge 4$, and $m = 2, \dots, \lfloor \frac{p}{2} \rfloor$, where $0 \le i_1 < \cdots < i_m \le p - 1$ are the indexes such that $\alpha_{i_1} = \cdots = \alpha_{i_m} = 1$.

In addition, if $p \ge 2$, $m = 1, \dots, \lfloor \frac{p}{2} \rfloor$ and $\alpha \in \Lambda_p^m$, let $0 \le i_1 < \cdots < i_m \le p - 1$ be the indexes such that $\alpha_{i_1} = \cdots = \alpha_{i_m} = 1$. Then, we define the binary multi-index $\bar{\alpha}$ of order p as

$$\bar{\alpha}_{i_j} = \bar{\alpha}_{i_j+1} = 0, \quad j = 1, \dots, m, \ \text{ and } \bar{\alpha}_i = 1 \ \text{ otherwise,}$$

where if $i_m = p - 1$, then $\bar{\alpha}_{p-1} = \bar{\alpha}_0 = 0$. Moreover, if $\alpha = (0, \dots, 0)$; that is, if $\alpha \in \Lambda_p^0$, then we define $\bar{\alpha} = (1, \dots, 1)$. It is clear that, in any case, $|\bar{\alpha}| = p - 2m$.

Lemma 3 ([9, Theorem 4.4]) *Given $n \in \mathbb{N}^*$ and $1 \le p < n$, the Floquet function $q_p \colon \mathbb{A}_p^{n+2} \times (\mathbb{R}^+)_p^{n+1} \longrightarrow \mathbb{R}$ is determined by*

$$q_p(\mathsf{a}, \mathsf{b}) = \frac{1}{2} \left(\prod_{k=0}^{p-1} b_k \right)^{-1} \sum_{j=0}^{\lfloor \frac{p}{2} \rfloor} (-1)^j \sum_{\alpha \in \Lambda_p^j} \prod_{k=0}^{p-1} b_k^{2\alpha_k} a_{k+1}^{\bar{\alpha}_k}.$$

Observe that when $p = 1$, $b_0 = \cdots = b_n = b > 0$ and $a_1 = \cdots = a_n = a$, then $\Lambda_1^0 = \{0\}$, so $\alpha_0 = 0$, $\bar{\alpha}_0 = 1$ and $q_1(\mathsf{a}, \mathsf{b}) = \dfrac{a}{2b}$. Moreover, since for any $\alpha \in \Lambda_p^j$, $j = 0, \dots, \lfloor \frac{p}{2} \rfloor$, the function $\prod_{k=0}^{p-1} b_k^{2\alpha_k} (a_{k+1} - x)^{\bar{\alpha}_k}$ is a polynomial of degree $p - 2j$, we conclude that $q_p(\mathsf{a} - x\mathsf{e}, \mathsf{b})$ is a polinomial in $x \in \mathbb{R}$ with degree equal to p and main coefficient $\dfrac{1}{2}\Big(\prod_{k=0}^{p-1} b_k \Big)^{-1}$.

Clearly, the main difficulty to compute $q_p(\mathsf{a} - x\mathsf{e}, \mathsf{b})$ is the obtention of the binary multi–indexes involved in them. In general, this is a difficult task and, in fact the number of multi–indexes in Λ_p^j, $0 \le j \le \lfloor \frac{p}{2} \rfloor$, grows dramatically with p, see [9, Proposition 4.2].

As a straightforward consequence of Lemmas 2 and 3, we can obtain the spectral information for Jacobi almost p-Toeplitz matrices.

For $\mathsf{a} \in \mathbb{A}_p^{n+2}$, we consider its periodicity values $\sigma = a_0 - a_p$ and $\tau = a_{n+1} - a_r$. Therefore, $a_0 = a_p + \sigma$ and $a_{n+1} = a_r + \tau$ and in particular, when $r = 0$ then

$a_{n+1} = a_p + \sigma + \tau$. Moreover, $\mathsf{a} \in \mathbb{R}_p^{n+2}$ iff $\sigma = \tau = 0$. In addition, for $\mathsf{b} \in (\mathbb{R}_+^{n+1})_p$ we have that

$$\mathsf{D}(\mathsf{b}) = b_1 \cdots b_n = b_0^{-1} \left(\prod_{j=0}^{r-1} b_j \right) \left(\prod_{j=0}^{p-1} b_j^m \right).$$

For $\mathsf{a} \in \mathbb{A}_p^{n+2}$ and $\mathsf{b} \in (\mathbb{R}^+)_p^{n+1}$ we define the *associated fundamental polynomials* as $v_k(x)$, $k = 0, \ldots, 2p-1$, recurrently given as

$$\begin{aligned}
v_0(x) &= b_0, \\
v_1(x) &= a_0 - x = a_p + \sigma - x, \\
v_k(x) &= b_{k-1}^{-1}\big((a_{k-1} - x)v_{k-1}(x) - b_{k-2}v_{k-2}(x)\big), \qquad k = 2, \ldots, p, \qquad \text{if } p \geq 2, \\
v_{p+1}(x) &= b_0^{-1}\big((a_p - x)v_p(x) - b_{p-1}v_{p-1}(x)\big), \\
v_{p+k}(x) &= b_{k-1}^{-1}\big((a_{k-1} - x)v_{p+k-1}(x) - b_{k-2}v_{p+k-2}(x)\big), \; k = 2, \ldots, p-1, \text{ if } p \geq 3.
\end{aligned}$$

It is clear that, for any $x \in \mathbb{R}$, $v_0(x), v_1(x), \ldots, v_{2p-1}(x)$ are the first $2p$ entries of the fundamental solution of the Schrödinger equation $\Delta_{\mathsf{a}-x\mathsf{e},\mathsf{b}}(\mathsf{z}) = \mathsf{0}$ on $\{1, \ldots, n\}$. The remain entries of the fundamental solution could be obtained from the fundamental polynomials. For instance, we have that $v_{2p}(x) = 2q_p(\mathsf{a} - x\mathsf{e}, \mathsf{b})v_p(x) - b_0$.

In addition, we also define the *coefficient polynomials* as

$$\begin{aligned}
\Phi_0(x) &= (a_p + \sigma + \tau - x)v_p(x) - b_{p-1}v_{p-1}(x), \\
\Psi_0(x) &= b_0(a_p + \sigma + \tau - x) + b_{p-1}\big(v_{2p-1}(x) - 2q(x)v_{p-1}(x)\big),
\end{aligned}$$

and when $p \geq 2$, for any $1 \leq r < p$ as

$$\Phi_r(x) = \tau v_{p+r}(x) + b_r v_{p+r+1}(x) \text{ and } \Psi_r(x) = \tau v_r(x) + b_r v_{r+1}(x).$$

Our main result is the following one, that represents an adaptation of the general claim in Theorem 1 to this specific case.

Theorem 2 *For any $\mathsf{a} \in \mathbb{A}_p^{n+2}$ and $\mathsf{b} \in (\mathbb{R}_+^{n+1})_p$ consider $v_0(x), \ldots, v_{2p-1}(x)$ and $\Phi_r(x)$, $\Psi_r(x)$ their associated fundamental and coefficient polynomials, respectively. Then, the characteristic polynomial of the Jacobi almost p-Toeplitz matrix $\mathsf{J}(\mathsf{a}, \mathsf{b})$ is*

$$P_{\mathsf{a},\mathsf{b}}(x) = \mathsf{D}(\mathsf{b})\Big[\Phi_r(x)U_{m-1}\big(q_p(\mathsf{a} - x\mathsf{e}, \mathsf{b})\big) - \Psi_r(x)U_{m-2}\big(q_p(\mathsf{a} - x\mathsf{e}, \mathsf{b})\big)\Big].$$

Therefore, $P_{\mathsf{a},\mathsf{b}}(\lambda)$ has $n+2$ different roots and when $P_{\mathsf{a},\mathsf{b}}(\lambda) = 0$, then $\mathcal{V}_{\mathsf{a},\mathsf{b}}(\lambda) =$ span$\big\{\mathsf{v}(\lambda)\big\}$, where the entries of $\mathsf{v}(\lambda)$ are given by

$$v_{kp+j}(\lambda) = v_{p+j}(\lambda)U_{k-1}\big(q_p(\mathsf{a} - \lambda\mathsf{e}, \mathsf{b})\big) - v_j(\lambda)U_{k-2}\big(q_p(\mathsf{a} - \lambda\mathsf{e}, \mathsf{b})\big)$$

for any $k = 0, \dots, m$, *if* $j = 0, \dots, r$, *and any* $k = 0, \dots, m-1$, *if* $j = r+1, \dots, p-1$.

The result in Theorem 2 is quite general and still far from give us manageable expressions for the spectral characteristics of the Jacobi almost-Toeplitz matrices. For instance, finding the roots of the polynomial $P_{\mathsf{a},\mathsf{b}}(x)$ is usually a numerical challenge, since explicit expressions would be available only in some special cases. We end this section describing a specific case in which we can go further, although we are still in a fairly general scenario.

When $\mathsf{a} \in \mathbb{A}_p^{n+2}$, $\mathsf{b} \in (\mathbb{R}_+^{n+1})_p$ and $n+1 = mp + r$ with $0 \le r < p$, we say the pair (a, b) is *β-compatible* if there exists $\beta \in \mathbb{R}$ such that

$$\Phi_r(x) = \big(2q_p(\mathsf{a} - x\mathsf{e}, \mathsf{b}) + \beta\big)\Psi_r(x). \tag{10}$$

Observe that when $r \ge 1$ the coefficients polynomials Φ_r and Ψ_r have degree $p+r+1$ and $r+1$ respectively, whereas when $r = 0$, Φ_0 has degree $p+1$ and Ψ_0 has degree less or equal to $2p-1$. Therefore, when $r = 0$, the pair (a, b) only can be β-compatible for some β, when the polynomial $v_{2p-1}(x) - 2q_p(\mathsf{a} - x\mathsf{e}, \mathsf{b})v_{p-1}(x)$ has degree less that or equal to 1 and $\Psi_0(x)$ has degree equal to 1.

As we can see below, when the pair (a, b) is compatible, the characteristic polynomial of $\mathsf{J}(\mathsf{a}, \mathsf{b})$ is a multiple of the polynomial

$$U_m\big(q_p(\mathsf{a} - x\mathsf{e}, \mathsf{b})\big) + \beta U_{m-1}\big(q_p(\mathsf{a} - x\mathsf{e}, \mathsf{b})\big),$$

so the eigenvalues of $\mathsf{J}(\mathsf{a}, \mathsf{b})$ are related with the zeroes of Chebyshev polynomials. Specifically, let $z_{m,1}(\beta) > \cdots > z_{m,m}(\beta)$ be the roots of the Chebyshev polynomial $U_{m,\beta}(x) = U_m(x) + \beta U_{m-1}(x)$ and $\theta_{m,1}(\beta), \dots, \theta_{m,m}(\beta)$ the arguments given in Proposition 4, see the Appendix (Sect. 6).

Theorem 3 *Assume that* $\mathsf{a} \in \mathbb{A}_p^{n+2}$ *and* $\mathsf{b} \in (\mathbb{R}_+^{n+1})_p$ *are β-compatible and consider* $\sigma, \tau \in \mathbb{R}$ *and* $v_0(x), \dots, v_{2p-1}(x)$ *be their periodicity values and their associated fundamental polynomials, respectively. Consider also the coefficient polynomial*

$$\Psi_r(x) = \begin{cases} b_0(a_p + \sigma + \tau - x) + b_{p-1}\big(v_{2p-1}(x) - 2q(x)v_{p-1}(x)\big), & \text{when } r = 0,\\ \tau v_r(x) + b_r v_{r+1}(x), & \text{when } 1 \le r < p. \end{cases}$$

Then, the characteristic polynomial of the Jacobi almost p-Toeplitz matrix $\mathsf{J}(\mathsf{a}, \mathsf{b})$ *is*

$$P_{\mathsf{a},\mathsf{b}}(x) = D(\mathsf{b})\Psi_r(x)U_{m,\beta}\big(q_p(\mathsf{a} - x\mathsf{e}, \mathsf{b})\big)$$

and hence, the following properties hold:

1. $\Psi_r(x)$ *has degree* $r+1$ *and* $r+1$ *simple real roots,* $\lambda_{mp+1}, \dots, \lambda_{n+2}$.
2. *For any* $j = 1, \dots, m$, *the equation* $q_p(\mathsf{a} - x\mathsf{e}, \mathsf{b}) = z_{m,j}(\beta)$ *has p different real solutions* $\lambda_{(j-1)p+1}, \dots, \lambda_{jp}$.
3. $\{\lambda_k\}_{k=1}^{n+2}$ *are the eigenvalues of* $\mathsf{J}(\mathsf{a}, \mathsf{b})$.

Moreover, when $|\beta| \le 1 + \frac{1}{m}$*, then for any* $j = 1, \ldots, m$*, and any* $\ell = 1, \ldots, p$*, we have that* $\mathcal{V}_{\mathsf{a},\mathsf{b}}(\lambda_{(j-1)p+\ell}) = \operatorname{span}\{\mathsf{v}(\lambda_{(j-1)p+\ell})\}$*, where the entries of* $\mathsf{v}(\lambda_{(j-1)p+\ell})$ *are given by*

$$\begin{aligned} v_{kp+i}(\lambda_{(j-1)p+\ell}) &= v_{p+i}(\lambda_{(j-1)p+\ell}) \sin\big(k\theta_{m,j}(\beta)\big) \\ &\quad - v_i(\lambda_{(j-1)p+\ell}) \sin\big((k-1)\theta_{m,j}(\beta)\big), \end{aligned}$$

for any $k = 0, \ldots, m$*, if* $i = 0, \ldots, r$*, and any* $k = 0, \ldots, m-1$*, if* $i = r+1, \ldots, p-1$*.*

When $|\beta| > 1 + \frac{1}{m}$*, the eigenvectors are the same except* $\mathsf{v}(\lambda_j)$*,* $j = 1, \ldots, p$*, for* $\beta < 0$*, or* $\mathsf{v}(\lambda_{(m-1)p+j})$*,* $j = 1, \ldots, p$*, for* $\beta > 0$*, for which their entries are respectively given by*

$$\begin{aligned} v_{kp+i}(\lambda_j) &= v_{p+i}(\lambda_j) \sinh\big(k\theta_{m,1}(\beta)\big) - v_i(\lambda_j) \sinh\big((k-1)\theta_{m,j}(\beta)\big), \\ v_{kp+i}(\lambda_{(j-1)p+\ell}) &= (-1)^k\big[v_{p+i}(\lambda_{(j-1)p+\ell}) \sinh\big(k\theta_{m,m}(\beta)\big) \\ &\quad + v_i(\lambda_{(j-1)p+\ell}) \sinh\big((k-1)\theta_{m,m}(\beta)\big)\big]. \end{aligned}$$

for any $k = 0, \ldots, m$*, if* $i = 0, \ldots, r$*, and any* $k = 0, \ldots, m-1$*, if* $i = r+1, \ldots, p-1$*.*

Remark 5 In the conditions of Theorem 3, the eigenvectors corresponding to the last eigenvalues $\lambda_{mp+1}, \ldots, \lambda_{n+2}$ do not have, in general, another expression that the one given in Theorem 2.

Even in the case in which the coefficients a and b are β-compatible, it is difficult to obtain a closed formula for the eigenvalues and eigenvectors of $\mathsf{J}(\mathsf{a}, \mathsf{b})$. In the rest of this paper, we will be interested in the subclass of matrices of the Jacobi almost-Toeplitz matrices that are also bisymmetric. This condition not only determines the uniqueness of the solution for the inverse eigenvalue problem, but also introduces conditions on the coefficients that allow us to simplify the previous general expressions. In addition, we use the notations in Theorems 2 and 3 and hence $\{\lambda_j\}_{j=1}^{n+2}$ denote the $n+2$ eigenvalues of $\mathsf{J}(\mathsf{a}, \mathsf{b})$ and for any $j = 1, \ldots, n+2$, $\mathsf{v}(\lambda_j)$ is an eigenvector corresponding to λ_j.

Although it is possible to describe the structure of vectors $\mathsf{a} \in \mathbb{S}^{n+1} \cap \mathbb{A}_p^{n+2}$ and $\mathsf{b} \in \mathbb{S}^{n+1} \cap (\mathbb{R}_+^{n+1})_p$, where $n+1 = mp + r$ with $0 \le r < p$, we limite ourselves to obtain such conditions in each analizad case. In any situation, we determine the fundamental and the coefficients polynomials and also the Floquet constant $q_p(\mathsf{a} - x\mathsf{e}, \mathsf{b})$ for any $x \in \mathbb{R}$.

Although our developments work for general periods, for reasons of space and clarity, we will restrict ourselves to periods not exceeding 2. In all cases, we also study under which conditions the pair (a, b) is β-compatible. The obtained results are enough to capture all the explicit formulas known so far, showing the unifying nature of our methodology.

4 Bisymmetric Jacobi Almost 1-Toeplitz Matrices

We begin our study of bisymmetric Jacobi almost-Toeplitz matrices for the easiest and most well-known case, when the period is equal to 1. As we have already mentioned in the previous section, this means that the coefficients of the corresponding homogeneous Schödinger equation (7) are constant. Since $p = 1$, we have that $n + 1 = mp + r$ where $m = n + 1$ and $r = 0$. Then, $\mathsf{a} \in \mathbb{S}^{n+1} \cap \mathbb{A}_1^{n+2}$ and $\mathsf{b} \in \mathbb{S}^{n+1} \cap (\mathbb{R}_+^{n+1})_1$ iff there exist $a, b, \sigma \in \mathbb{R}$ with $b > 0$ such that $\mathsf{a} = (a + \sigma, a, \ldots, a, a + \sigma)$ and $\mathsf{b} = b\mathsf{e}$. Then, $\mathsf{J}(\mathsf{a}, \mathsf{b})$ is a Jacobi 1-Toeplitz matrix iff $\sigma = 0$.

In this scenario, $D(\mathsf{b}) = b^n$, $q_1(\mathsf{a} - x\mathsf{e}, \mathsf{b}) = \dfrac{a - x}{2b}$, $v_0(x) = b$, $v_1(x) = a + \sigma - x$, and the coefficient polynomials are

$$\begin{aligned}\Phi_0(x) &= (a + \sigma - x)^2 - b^2 = (a + 2\sigma - x)(a - x) + \sigma^2 - b^2,\\ \Psi_0(x) &= b(a + \sigma - x) + b\big(a + \sigma - x - (a - x)\big) = b(a + 2\sigma - x).\end{aligned}$$

Therefore, $\Phi_0(x) = 2q_1(\mathsf{a} - x\mathsf{e}, \mathsf{b})\Psi_0(x) + \sigma^2 - b^2$, and hence the pair (a, b) is β-compatible iff $\sigma = \pm b$ and then $\beta = 0$.

Proposition 1 *Given $a, \sigma \in \mathbb{R}$ and $b > 0$, consider $\mathsf{a} = (a + \sigma, a, \ldots, a, a + \sigma)$ and $\mathsf{b} = b\mathsf{e}$. Then, the characteristic polynomial of the bisymmetric Jacobi almost 1-Toeplitz matrix $\mathsf{J}(\mathsf{a}, \mathsf{b})$ is the Chebyshev polynomial*

$$P_{\mathsf{a},\mathsf{b}}(x) = b^n\Big[b(a + 2\sigma - x)U_{n+1}\Big(\frac{a - x}{2b}\Big) + (\sigma^2 - b^2)U_n\Big(\frac{a - x}{2b}\Big)\Big].$$

Therefore, $P_{\mathsf{a},\mathsf{b}}$ has $n + 2$ different roots $\lambda_1 < \cdots < \lambda_{n+2}$ and for any $j = 1, \ldots, n + 1$, the entries of $\mathsf{v}(\lambda_j) \in \mathbb{R}^{n+2}$ are

$$v_k(\lambda_j) = bU_k\Big(\frac{a - \lambda_j}{2b}\Big) + \sigma U_{k-1}\Big(\frac{a - \lambda_j}{2b}\Big),\ k = 0, \ldots, n + 1.$$

Proof From Theorem 2, if $q(x) = \frac{a-x}{2b}$, the characteristic polynomial of $\mathsf{J}_{a,b,\sigma}$ is

$$\begin{aligned}P_{\mathsf{a},\mathsf{b}}(x) &= b^n\Big[\big(2q(x)\Psi_0(x) + \sigma^2 - b^2\big)U_n\big(q(x)\big) - \Psi_0(x)U_{n-1}\big(q(x)\big)\Big]\\ &= b^n\Big[\Psi_0(x)U_{n+1}\big(q(x)\big) + (\sigma^2 - b^2)U_n\big(q(x)\big)\Big].\end{aligned}$$

Finally, from Theorem 2 and taking into account the Chebyshev recurrence, we also have that for any $x \in \mathbb{R}$, the entries of $\mathsf{v}(x)$ are given by

$$\begin{aligned}v_k(x) &= (a + \sigma - x)U_{k-1}\big(q(x)\big) - bU_{k-2}\big(q(x)\big)\\ &= \big(\sigma + 2bq(x)\big)U_{k-1}\big(q(x)\big) - 2bq(x)bU_{k-1}\big(q(x)\big) + bU_k\big(q(x)\big)\\ &= bU_k\big(q(x)\big) + \sigma U_{k-1}\big(q(x)\big),\end{aligned}$$

$k = 0, \ldots, n + 1$. □

The results in Proposition 1 are good examples of how complicated is the computation of the eigenvalues, eigenvectors, and also the determinant of Jacobi matrices even when the matrix has a nice and simple structure. Although we have an explicit expression of the characteristic polynomial and the eigenvectors, in general, determining the eigenvalues may still be far from our reach, at least by elementary methods.

In the next result, we describe the well-known case of Jacobi 1-Toeplitz matrices.

Corollary 1 *Given $a \in \mathbb{R}$ and $b > 0$ the eigenvalues of the Jacobi 1-Toeplitz matrix $\mathsf{J}(a\mathsf{e}, b\mathsf{e})$ are $\lambda_j = a - 2b\cos\left(\frac{j\pi}{n+3}\right)$, $j = 1, \dots, n+2$. In addition for any $j = 1, \dots, n+2$, $\mathsf{y}^j \in \mathbb{R}^{n+2}$ whose entries are $y_k^j = \sin\left(\frac{kj\pi}{n+3}\right)$, $k = 1, \dots, n+2$ is an eigenvector of $\mathsf{J}(a\mathsf{e}, b\mathsf{e})$ corresponding to λ_j. Moreover, it is satisfied that $\det(\mathsf{J}(a\mathsf{e}, b\mathsf{e}) = b^{n+2} U_{n+2}\left(\frac{a}{2b}\right)$.*

Proof In this case, the characteristic polynomial is $P_{a\mathsf{e},b\mathsf{e}}(x) = b^{n+2} U_{n+2}\left(\frac{a-x}{2b}\right)$ and hence $P_{a\mathsf{e},b\mathsf{e}}(x) = 0$ iff $\frac{a-x}{2b} = \cos\left(\frac{j\pi}{n+3}\right)$ for some $j = 1, \dots, n+2$. In addition, for any $k = 1, \dots, n+2$, $v_k(\lambda_j) = bU_k\big(\cos\left(\frac{j\pi}{n+3}\right)\big) = b\big(\sin\left(\frac{j\pi}{n+3}\right)\big)^{-1} y_k^j$. □

The results in the previous Corollary are well-known, see [19, Example 7.4], [5, Proposition 3.9] and also [26, page 514]. We remark that in this two last references, second order difference equations techniques are considered, but in the last one using *Binet basis* instead the Chebyshev one; that is, considering the two roots of the polynomial $-bx^2 + 2qx - b$.

Remark 6 In Corollary 1, we have that a is an eigenvalue of $\mathsf{J}(a\mathsf{e}, b\mathsf{e})$ iff n is odd and then $\lambda_{m+1} = a$, where $m = \frac{n+1}{2}$ and $\mathsf{y}^{m+1} = \big(1, 0, -1, 0, \dots, 0, (-1)^m\big)$.

The following result corresponds to the remained compatible cases; that is when $\sigma^2 = b^2$.

Corollary 2 *For any $a, b \in \mathbb{R}$ where $b > 0$ consider the vectors*

$$\mathsf{a}_b^+ = (a+b, a, \dots, a, a+b) \text{ and } \mathsf{a}_b^- = (a-b, a, \dots, a, a-b).$$

Then , the following results hold:

(i) The eigenvalues of the Jacobi matrix $\mathsf{J}(\mathsf{a}_b^+, b\mathsf{e})$ are $\lambda_j = a - 2b\cos\left(\frac{j\pi}{n+2}\right)$, $j = 1, \dots, n+2$. In addition, for any $j = 1, \dots, n+1$, the vector $\mathsf{y}^j \in \mathbb{R}^{n+2}$ whose entries are $y_k^j = \sin\left(\frac{(2k+1)j\pi}{2n+4}\right)$, $k = 0, \dots, n+1$, is an eigenvector of $\mathsf{J}(\mathsf{a}_b^+, b\mathsf{e})$ corresponding to λ_j, whereas y^{n+2} whose entries are $y_k^{n+2} = (-1)^k$, $k = 0, \dots, n+1$, is an eigenvector of $\mathsf{J}(\mathsf{a}_b^+, b\mathsf{e})$ corresponding to $\lambda_{n+2} = a + 2b$.

(ii) *The eigenvalues of the Jacobi matrix* $\mathsf{J}(\mathsf{a}_b^-, b\mathsf{e})$ *are* $\lambda_j = a - 2b\cos\Big(\dfrac{j\pi}{n+2}\Big)$, $j = 0, \dots, n+1$. *In addition,* z^0 *whose entries are* $z_k^0 = 1$, $k = 0, \dots, n+1$, *is an eigenvector of* $\mathsf{J}(\mathsf{a}_b^-, b\mathsf{e})$ *corresponding to* $\lambda_0 = a - 2b$, *whereas for any* $j = 1, \dots, n+1$, *the vector* $\mathsf{z}^j \in \mathbb{R}^{n+2}$ *whose entries are* $z_k^j = \cos\big(\frac{(2k+1)j\pi}{2n+4}\big)$, $k = 0, \dots, n+1$, *is an eigenvector of* $\mathsf{J}(\mathsf{a}_b^-, b\mathsf{e})$ *corresponding to* λ_j.

Moreover, $\det\big(\mathsf{J}(\mathsf{a}_b^\pm, b\mathsf{e})\big) = b^{n+1}(a \pm 2b)U_{n+1}\big(\frac{a}{2b}\big)$.

Proof In this case, $\Psi_0(x) = b(a \pm 2b - x)$, so the characteristic polynomial is

$$P_{\mathsf{a},\mathsf{b}}(x) = b^{n+1}(a \pm 2b - x)U_{n+1}\Big(\frac{a-x}{2b}\Big)$$

and hence $P_{\mathsf{a},\mathsf{b}}(\lambda) = 0$ iff either $\lambda = a \pm 2b$ or $\frac{a-\lambda}{2b} = \cos\big(\frac{j\pi}{n+2}\big)$ for some $j = 1, \dots, n+1$. The statements follow in a similar way than the previous Corollary. □

The expression for the eigenvalues of $\mathsf{J}(\mathsf{a}_b^\pm, b\mathsf{e})$ appears recurrently in the literature, see for instance [19, Example 7.7] and [31, Theorem 4], and also [30].

All the three cases in the above corollaries were considered in [7, Identity (25)] and also in [8, Theorem 2], where the main motivation was the discretization of some boundary value problems as an intermediate step to study Navier-Stokes equations with non-slip boundary conditions. In addition, the particular choice $a = 0$ for the three cases was studied in [7, Identity (25)].

5 Bisymmetric Jacobi Almost 2-Toeplitz Matrices

First of all, we have that $\Lambda_2^0 = \{(0,0)\}$, $\Lambda_2^1 = \{(1,0), (0,1)\}$, $\bar{\Lambda}_2^0 = \{(1,1)\}$ and $\bar{\Lambda}_2^1 = \{(0,0), (0,0)\}$. Therefore, applying Lemma 3 for any $\mathsf{a} \in \mathbb{A}_p^{n+2}$, $\mathsf{b} \in \mathbb{R}_p^{n+1}$ and $x \in \mathbb{R}$, we get that $q(x) = q_2(\mathsf{a} - x\mathsf{e}, \mathsf{b}) = \frac{1}{2b_0b_1}\big((a_1 - x)(a_2 - x) - b_0^2 - b_1^2\big)$. Moreover, the fundamental polynomials are $v_0(x) = b_0$, $v_1(x) = a_2 + \sigma - x$, and

$$\begin{aligned} v_2(x) &= 2b_0q(x) + b_1 + b_1^{-1}\sigma(a_1 - x) = 2q(x)v_0 + b_1 + b_1^{-1}\sigma(a_1 - x),\\ v_3(x) &= 2q(x)(a_2 + \sigma - x) + b_0b_1^{-1}\sigma = 2q(x)v_1(x) + b_0b_1^{-1}\sigma. \end{aligned}$$

If $m = \lfloor\frac{n+1}{2}\rfloor$, then $n+1 = 2m+1$ when n is even and $n+1 = 2m$ when n is odd.

5.1 *Case n Even*

If $n = 2m$, $m \geq 1$, then $\mathsf{a} \in \mathbb{A}_2^{n+2}$, $\mathsf{b} \in (\mathbb{R}_+^{n+1})_2$ iff there exist $a_1, a_2, b_0, b_1, \sigma, \tau \in \mathbb{R}$ such that $\mathsf{a} = (a_2 + \sigma, a_1, a_2, \dots, a_1, a_2, a_1 + \tau)$ and $\mathsf{b} = (b_0, b_1, \dots, b_0, b_1, b_0)$, and hence $\mathsf{b} \in \mathbb{S}^{n+1} \cap (\mathbb{R}^{n+1}_+)_2$, whereas $\mathsf{a} \in \mathbb{S}^{n+2} \cap \mathbb{A}^{n+2}_2$ iff $a_2 = a_1$ and moreover $\tau = \sigma$. Therefore, in this paragraph we consider $a, \sigma \in \mathbb{R}$, $b_0, b_1 > 0$ and the vectors $\mathsf{a} = (a+\sigma, a, \dots, a, a+\sigma) \in \mathbb{R}^{n+2}$ and $\mathsf{b} = (b_0, b_1, \dots, b_0, b_1, b_0) \in \mathbb{R}^{n+1}$.

Under these hypotheses, $D(\mathsf{b}) = (b_0 b_1)^m$, $q(x) = \frac{(a-x)^2}{2b_0b_1} - \cosh\big(\ln(b_0^{-1}b_1)\big)$, since $b_0^2 + b_1^2 = 2b_0b_1\cosh\big(\ln(b_0^{-1}b_1)\big)$, and the coefficient polynomials are

$$\begin{aligned}\Psi_1(x) &= \sigma(a+\sigma-x) + 2b_0b_1q(x) + b_1^2 + \sigma(a-x) = (a+\sigma-x)^2 - b_0^2,\\ \Phi_1(x) &= 2q(x)\big((a+\sigma-x)^2 - b_0^2\big) + b_0b_1^{-1}(\sigma^2 - b_1^2)\\ &= 2q(x)\Psi_1(x) + b_0b_1^{-1}(\sigma^2 - b_1^2).\end{aligned}$$

Proposition 2 *The characteristic polynomial of the bisymmetric Jacobi almost 2-Toeplitz matrix* $\mathsf{J}(\mathsf{a},\mathsf{b})$ *is*

$$P_{\mathsf{a},\mathsf{b}}(x) = b_0^m b_1^{m-1}\Big[b_1\big((a+\sigma-x)^2 - b_0^2\big)U_m\big(q(x)\big) + b_0(\sigma^2 - b_1^2)U_{m-1}\big(q(x)\big)\Big].$$

Therefore, $P_{\mathsf{a},\mathsf{b}}$ *has* $n+2$ *roots* $\lambda_1 < \cdots < \lambda_{n+2}$ *and for any* $j = 1, \dots, n+2$, *the entries of* $\mathsf{v}(\lambda_j)$ *are*

$$\begin{aligned}v_{2k}(\lambda_j) &= b_1b_0U_k\big(q(\lambda_j)\big) + \big(b_1^2 + \sigma(a-\lambda_j)\big)U_{k-1}\big(q(\lambda_j)\big), \quad k = 0, \dots, m,\\ v_{2k+1}(\lambda_j) &= b_1(a+\sigma-\lambda_j)U_k\big(q(\lambda_j)\big) + b_0\sigma U_{k-1}\big(q(\lambda_j)\big), \quad k = 0, \dots, m.\end{aligned}$$

Proof From Theorem 2 and using that $U_k\big(q(x)\big)$ satisfies the Chebyshev recurrence, we have that

$$P_{\mathsf{a},\mathsf{b}}(x) = (b_0b_1)^m\Big[\big((a+\sigma-x)^2 - b_0^2\big)U_m\big(q(x)\big) + b_0b_1^{-1}(\sigma^2 - b_1^2)U_{m-1}\big(q(x)\big)\Big].$$

When $P_{\mathsf{a},\mathsf{b}}(\lambda) = 0$, from the last part of Theorem 2, and using newly that $U_k\big(q(x)\big)$ satisfies the Chebyshev recurrence, for any $k = 0, \dots, m$ we obtain the components of the eigenvectors. □

Remark 7 When $b_1 = b_0 = b$ in Proposition 2, we newly recover the Jacobi almost 1-Toeplitz case analyzed in Proposition 1: Since $q(x) = 2\big(\frac{a-x}{2b}\big)^2 - 1$, we can obtain the expression for the characteristic polynomial from Identities (12) in the Appendix. The same applies to the expression for the corresponding eigenvectors.

Since $\Phi_1(x) = 2q(x)\Psi_1(x) + b_0b_1^{-1}(\sigma^2 - b_1^2)$, the pair (a,b) is β-compatible iff $\beta = 0$ and this happens iff $\sigma = \pm b_1$. Therefore, the compatible cases correspond to consider either $\mathsf{a}^+_{b_1} = (a+b_1, a, \dots, a, a+b_1)$ or $\mathsf{a}^-_{b_1} = (a-b_1, a, \dots, a, a-b_1)$.

Corollary 3 *The characteristic polynomial of* $\mathsf{J}(\mathsf{a}^\pm_{b_1}, \mathsf{b})$ *is*

$$P_{\mathsf{a}^\pm_{b_1},\mathsf{b}}(x) = (b_0b_1)^m\big((a \pm b_1 - x)^2 - b_0^2\big)U_m\big(q(x)\big).$$

Moreover, considering the positive real values $\eta_j = \sqrt{(b_0 - b_1)^2 + 4b_0 b_1 \cos^2\left(\frac{j\pi}{n+2}\right)}$, $j = 1, \dots, m$, *the following results hold:*

(i) *The eigenvalues of* $\mathsf{J}(\mathsf{a}^+_{b_1}, \mathsf{b})$ *are*

$$a - \mu_1 < \cdots < a - \mu_m < a - b_0 + b_1 < a + \mu_m < \cdots < a + \mu_1 < a + b_0 + b_1.$$

Moreover, $\mathsf{y}^{m+1} = (1, 1, -1, -1, \dots, (-1)^m, (-1)^m)$ *and* $\mathsf{y}^{n+2} = (1, -1, \dots, 1, -1)$ *are eigenvectors corresponding to* $\lambda_{m+1} = a - b_0 + b_1$ *and* $\lambda_{n+2} = a + b_0 + b_1$, *respectively, whereas for any* $j = 1, \dots, m$ *the vectors* $\mathsf{y}^j, \mathsf{y}^{m+1+j}$ *whose entries are*

$$\begin{aligned}
y^j_{2k} &= b_0 \sin\left(\tfrac{(k+1)j\pi}{m+1}\right) + (b_1 + \mu_j)\sin\left(\tfrac{kj\pi}{m+1}\right), && k = 0, \dots, m,\\
y^j_{2k+1} &= (b_1 + \mu_j)\sin\left(\tfrac{(k+1)j\pi}{m+1}\right) + b_0 \sin\left(\tfrac{j\pi}{m+1}\right), && k = 0, \dots, m,\\
y^{m+1+j}_{2k} &= b_0 \sin\left(\tfrac{(k+1)j\pi}{m+1}\right) + (b_1 - \mu_{m+1-j})\sin\left(\tfrac{kj\pi}{m+1}\right), && k = 0, \dots, m,\\
y^{m+1+j}_{2k+1} &= (b_1 - \mu_{m+1-j})\sin\left(\tfrac{(k+1)j\pi}{m+1}\right) + b_0 \sin\left(\tfrac{j\pi}{m+1}\right), && k = 0, \dots, m,
\end{aligned}$$

are eigenvectors for $\lambda_j = a - \mu_j$ *and* $\lambda_{m+1+j} = a + \mu_{m+1-j}$, *respectively.*

(ii) *The eigenvalues of* $\mathsf{J}(\mathsf{a}^-_{b_1}, \mathsf{b})$ *are*

$$a - b_0 - b_1 < a - \mu_1 < \cdots < a - \mu_m < a + b_0 - b_1 < a + \mu_m < \cdots < a + \mu_1.$$

Moreover, $\mathsf{z}^{m+1} = \mathsf{e}$ *and* $\mathsf{z}^{n+2} = \left(1, -1, -1, 1, \dots, (-1)^m, (-1)^{m+1}\right)$ *are eigenvectors corresponding to* $\lambda_1 = a - b_0 - b_1$ *and* $\lambda_{m+1} = a + b_0 - b_1$, *respectively, whereas for any* $j = 1, \dots, m$ *the vectors* $\mathsf{z}^{j+1}, \mathsf{z}^{m+2+j}$ *whose entries are*

$$\begin{aligned}
z^j_{2k} &= b_0 \sin\left(\tfrac{(k+1)j\pi}{m+1}\right) + (b_1 - \mu_j)\sin\left(\tfrac{kj\pi}{m+1}\right), && k = 0, \dots, m,\\
z^j_{2k+1} &= (b_1 - \mu_j)\sin\left(\tfrac{(k+1)j\pi}{m+1}\right) + b_0 \sin\left(\tfrac{j\pi}{m+1}\right), && k = 0, \dots, m,\\
z^{m+j}_{2k} &= b_0 \sin\left(\tfrac{(k+1)j\pi}{m+1}\right) + (b_1 + \mu_j)\sin\left(\tfrac{kj\pi}{m+1}\right), && k = 0, \dots, m,\\
z^{m+j}_{2k+1} &= (b_1 + \mu_j)\sin\left(\tfrac{(k+1)j\pi}{m+1}\right) + b_0 \sin\left(\tfrac{j\pi}{m+1}\right), && k = 0, \dots, m,
\end{aligned}$$

are eigenvectors for $\lambda_{j+1} = a - \mu_j$ *and* $\lambda_{m+2+j} = a + \mu_{m+1-j}$, *respectively.*

We end this paragraph analyzing the case of Jacobi 2-Toeplitz matrices, that corresponds to take $\sigma = 0$ and then $\mathsf{a} = a\mathsf{e}$. In addition, we also consider the Chebyshev polynomial $b_0 U_{m+1, b_0^{-1} b_1}(x) = b_0 U_{m+1}(x) + b_1 U_m(x)$. Then, applying part 2 of Proposition 4, for any $j = 1, \dots, m$, there exists $\frac{j\pi}{m+2} < \theta_j < \frac{(j+1)\pi}{m+2}$, such that $\cos(\theta_j)$ is a root of $b_0 U_{m+1, b_0^{-1} b_1}(x)$. Moreover, there exists $\theta_{m+1} > 0$ such that the following results hold:

(a) When $b_1 < b_0\left(1 + \frac{1}{m+1}\right)$, then $\frac{(m+1)\pi}{m+2} < \theta_{m+1} < \pi$ and $\cos\left(\theta_{m+1}\right)$ is a root of $b_0 U_{m+1, b_0^{-1} b_1}(x)$.

(b) When $b_1 = b_0\left(1 + \frac{1}{m+1}\right)$, then $\theta_{m+1} = \pi$ and -1 is a root of $b_0 U_{m+1, b_0^{-1} b_1}(x)$.

(c) When $b_1 > b_0\big(1+\frac{1}{m+1}\big)$, $\theta_{m+1} < \ln(b_0^{-1}b_1)$ and $-\cosh\big(\theta_{m+1}\big)$ is a root of $b_0 U_{m+1,b_0^{-1}b_1}(x)$.

In addition, for any $j = 1,\dots,m$, we define $\mu_j = \sqrt{\cosh\big(\ln(b_0^{-1}b_1)\big) + \cos(\theta_j)}$ and

$$\mu_{m+1} = \begin{cases} \sqrt{\cosh\big(\ln(b_0^{-1}b_1)\big) + \cos(\theta_{m+1})}, & \text{if} b_1 \le b_0\big(1+\frac{1}{m+1}\big), \\ \sqrt{\cosh\big(\ln(b_0^{-1}b_1)\big) - \cosh(\theta_{m+1})}, & \text{if } b_1 > b_0\big(1+\frac{1}{m+1}\big). \end{cases}$$

Observe that $\mu_j > 0,\ j = 1,\dots,m+1$ and that when $b_1 = b_0\big(1+\frac{1}{m+1}\big)$, then $\mu_{m+1} = \frac{b_0}{(m+1)\sqrt{2b_0b_1}}$.

Corollary 4 *The characteristic polynomial of the bisymmetric Jacobi 2-Toeplitz matrix* $\mathsf{J}(a\mathsf{e},\mathsf{b})$ *is*

$$P_{a\mathsf{e},\mathsf{b}}(x) = b_0^m b_1^{m+1}\Big[b_0 U_m\Big(\frac{(a-x)^2 - b_0^2 - b_1^2}{2b_0b_1}\Big) + b_1 U_{m-1}\Big(\frac{(a-x)^2 - b_0^2 - b_1^2}{2b_0b_1}\Big)\Big]$$

and hence the eigenvalues are $\lambda_j = a - \sqrt{2b_0b_1}\,\mu_j$ *and* $\lambda_{m+1+j} = a + \sqrt{2b_0b_1}\,\mu_j$, $j = 1,\dots,m+1$. *Moreover, when* $b_1 \le b_0\big(1+\frac{1}{m+1}\big)$, *for any* $j = 1,\dots,m+1$ *the vectors* $\mathsf{w}^j, \mathsf{w}^{m+1+j}$ *whose entries are*

$$\begin{aligned} w_{2k}^j = w_{2k}^{m+1+j} &= b_0 \sin\big((k+1)\theta_j\big) + b_1 \sin\big(k\theta_j\big), & k = 0,\dots,m, \\ v_{2k+1}^j &= \sqrt{2b_0b_1}\,\mu_j \sin\big((k+1)\theta_j\big), & k = 0,\dots,m, \\ v_{2k+1}^{m+1+j} &= -\sqrt{2b_0b_1}\,\mu_j \sin\big((k+1)\theta_j\big), & k = 0,\dots,m, \end{aligned}$$

are eigenvectors for λ_j *and* λ_{m+1+j}, *respectively. When* $b_1 > b_0\big(1+\frac{1}{m+1}\big)$, *for any* $\ell = 1,\dots,n+1$, $\ell \ne m+1$, *the vector* w^ℓ *is and eigenvector for* λ_ℓ, *whereas* $\hat{\mathsf{w}}^{m+1}$ *and* $\hat{\mathsf{w}}^{n+2}$ *whose entries are*

$$\begin{aligned} w_{2k}^{m+1} = w_{2k}^{n+2} &= b_0 \sinh\big((k+1)\theta_j\big) - b_1 \sinh\big(k\theta_j\big), & k = 0,\dots,m, \\ v_{2k+1}^{m+1} &= \sqrt{2b_0b_1}\,\mu_j \sinh\big((k+1)\theta_j\big), & k = 0,\dots,m, \\ v_{2k+1}^{n+2} &= -\sqrt{2b_0b_1}\,\mu_j \sinh\big((k+1)\theta_j\big), & k = 0,\dots,m, \end{aligned}$$

are eigenvectors for λ_{m+1} *and* λ_{nm+2}, *respectively.*

5.2 *Case n Odd*

If $n = 2m-1$, $m \ge 1$, then $\mathsf{a} \in \mathbb{A}_2^{n+2}$, $\mathsf{b} \in (\mathbb{R}_+^{n+1})_2$ iff there exist $a_1, a_2, b_0, b_1, \sigma, \tau \in \mathbb{R}$ such that $\mathsf{a} = (a_2+\sigma, a_1, a_2, \dots, a_1, a_2, a_1, a_2+\sigma+\tau)$ and $\mathsf{b} = (b_0, b_1, \dots, b_0, b_1)$, and hence, $\mathsf{a} \in \mathbb{S}^{n+2} \cap \mathbb{A}_2^{n+2}$ iff $\tau = 0$, whereas $\mathsf{b} \in \mathbb{S}^{n+1} \cap$

$\mathbb{R}^{n+1}_2$ iff $b_1 = b_0$. Therefore, in this paragraph we consider $a_1, a_2, \sigma \in \mathbb{R}$, $b > 0$ and the vectors $\mathsf{a} = (a_2 + \sigma, a_1, a_2, \dots, a_1, a_2, a_1, a_2 + \sigma) \in \mathbb{R}^{n+2}$ and $\mathsf{b} = b\mathsf{e} \in \mathbb{R}^{n+1}$.

Under these hypotheses, $D(\mathsf{b}) = b^n$, $q(x) = \dfrac{(a_1 - x)(a_2 - x)}{2b^2} - 1$ and the coefficient polynomials are

$$\begin{aligned}\Psi_0(x) &= bv_1(x) + b\big(v_3(x) - 2q(x)v_1(x)\big) = b(a_2 + 2\sigma - x),\\ \Phi_0(x) &= 2q(x)\Psi_0(x) + \sigma b^{-1}\big(\sigma(a_1 - x) + 2b^2\big).\end{aligned}$$

Proposition 3 *Under the above conditions, the characteristic polynomial of the bisymmetric Jacobi almost 2-Toeplitz matrix* $\mathsf{J}(\mathsf{a}, \mathsf{b})$ *is*

$$P_{\mathsf{a},\mathsf{b}}(x) = b^{n-1}\Big[b^2(a_2 + 2\sigma - x)U_m\big(q(x)\big) + \sigma\big(\sigma(a_1 - x) + 2b^2\big)U_{m-1}\big(q(x)\big)\Big].$$

Therefore, $P_{\mathsf{a},\mathsf{b}}$ *has* $n+2$ *roots* $\lambda_1 < \cdots < \lambda_{n+2}$ *and for any* $j = 1, \dots, n+2$, *the entries of* $\mathsf{v}(\lambda_j)$ *are*

$$\begin{aligned}v_{2k}(\lambda_j) &= bU_k\big(q(\lambda_j)\big) + \big(b + b^{-1}\sigma(a_1 - \lambda_j)\big)U_{k-1}\big(q(\lambda_j)\big), & k &= 0, \dots, m,\\ v_{2k+1}(\lambda_j) &= (a_2 + \sigma - \lambda_j)U_k\big(q(\lambda_j)\big) + \sigma U_{k-1}\big(q(\lambda_j)\big), & k &= 0, \dots, m-1.\end{aligned}$$

Proof It suffices to apply Theorem 2 and using that $U_k\big(q(x)\big)$ satisfies the Chebyshev recurrence as in Proposition 2. □

Remark 8 When $a_1 = a_2 = a$ in Proposition 3, we newly recover the Jacobi almost 1-Toeplitz case analyzed in Proposition 1: Since $q(x) = 2\big(\frac{a-x}{2b}\big)^2 - 1$, we can obtain the expression for the characteristic polynomial from Identities (12) of the Appendix. The same applies to the expression for the corresponding eigenvectors.

Next, we raised the question when the pair (a, b) is β-compatible.

Lemma 4 *The pair* (a, b) *is* β*-compatible iff* $\beta = b^{-2}\sigma^2$ *and either* $\sigma = 0$ *or* $\sigma = \sigma_\pm$, *where* $\sigma_\pm = \frac{1}{4}\Big(a_1 - a_2 \pm \sqrt{(a_1 - a_2)^2 + 16b^2}\Big)$. *Moreover, the following properties hold:*

(i) $\sigma_+ > 0$ *and* $\sigma_+^2 \le b^2\big(1 + \frac{1}{m}\big)$ *iff* $a_1 \le a_2 + \frac{2b}{\sqrt{m(m+1)}}$, *where the first inequality is an equality iff the second one is an equality,*
(i) $\sigma_- = -\frac{b^2}{\sigma_+} < 0$ *and* $\sigma_-^2 \le b^2\big(1 + \frac{1}{m}\big)$ *iff* $a_2 \le a_1 + \frac{2b}{\sqrt{m(m+1)}}$, *where the first inequality is an equality iff the second one is an equality.*

Proof The pair (a, b) is β-compatible iff for any $x \in \mathbb{R}$

$$\sigma\big(\sigma(a_1 - x) + 2b^2\big) = \beta b^2(a_2 + 2\sigma - x) = \beta b^2(a_1 - x + a_2 - a_1 + 2\sigma).$$

This happens iff $\beta b^2 = \sigma^2$ and moreover $\sigma\big(2\sigma^2 + \sigma(a_2 - a_1) - 2b^2\big) = 0$; that is, iff either $\sigma = 0$ or $\sigma = \frac{1}{4}\Big(a_1 - a_2 \pm \sqrt{(a_1 - a_2)^2 + 16b^2}\Big)$. In all cases, $\beta =$

$b^{-2}\sigma^2$. On the other hand, if we define the function $h\colon \mathbb{R} \longrightarrow \mathbb{R}$ as $h(x) = \frac{1}{8}\big(x^2 + x\sqrt{x^2+16b^2}\big)$, then $h(x) < 0$ when $x < 0$, $h(0) = 0$ and $h(x) > 0$ when $x > 0$. Moreover, $h(x) \le \frac{b^2}{m}$ iff either $x \le 0$ or $0 < x\sqrt{x^2+16b^2} \le \frac{8b^2}{m} - x^2$ when $x > 0$, which in turn is equivalent to $0 < x \le \frac{2b}{\sqrt{m(m+1)}}$. In addition, $h(x) = \frac{b^2}{m}$ iff $x = \frac{2b}{\sqrt{m(m+1)}}$. The proof ends by observing that $\sigma_+^2 = h(a_1 - a_2) + b^2$ whereas $\sigma_-^2 = h(a_2 - a_1) + b^2$. □

Let now σ satisfying the conditions in the above Lemma 4, that is, either $\sigma = 0$ or $\sigma = \sigma_\pm$. Then, from Proposition 3, the characteristic polynomial of the Jacobi almost 2-Toeplitz matrix $\mathsf{J}(\mathsf{a},\mathsf{b})$ is

$$P_{\mathsf{a},\mathsf{b}}(x) = b^{n+1}(a_2 + 2\sigma - x)\Big[U_m\big(q(x)\big) + b^{-2}\sigma^2 U_{m-1}\big(q(x)\big)\Big].$$

Corollary 5 *Let either* $\sigma = 0$ *or* $\sigma = \sigma_+$ *and* $a_1 \le a_2 + \frac{2b}{\sqrt{m(m+1)}}$ *or* $\sigma = \sigma_-$ *and* $a_2 \le a_1 + \frac{2b}{\sqrt{m(m+1)}}$. *For any* $j = 1,\dots,m$ *consider* $\theta_j = \frac{1}{2}\theta_{m,j}(b^{-2}\sigma^2)$ *given in Parts 1 and 2 of Proposition 4 and also* $\nu_j = \sqrt{(a_2-a_1)^2 + 16b^2\cos^2(\theta_j)}$. *Then the eigenvalues of* $\mathsf{J}(\mathsf{a},\mathsf{b})$ *are* $\lambda_{n+2} = a_2 + 2\sigma$,

$$\lambda_{2j-1} = \frac{1}{2}(a_1 + a_2 - \nu_j) \text{ and } \lambda_{2j} = \frac{1}{2}(a_1 + a_2 + \nu_j), \text{ for any } j = 1,\dots,m.$$

In addition, the vectors y^{2j-1} *and* y^{2j} *whose entries are*

$$\begin{aligned}
y_{2k}^{2j-1} &= 4b^2\cos(\theta_j)\sin\big((2k+1)\theta_j\big) \\
&\quad + \sigma(a_1 - a_2 + \nu_j)\sin(2k\theta_j), && k = 0,\dots,m,\\
y_{2k+1}^{2j-1} &= 4b\sigma\cos(\theta_j)\sin\big((2k+1)\theta_j\big) \\
&\quad + b(a_2 - a_1 + \nu_j)\sin\big(2(k+1)\theta_j\big), && k = 0,\dots,m-1,\\
y_{2k}^{2j} &= 4b^2\cos(\theta_j)\sin\big((2k+1)\theta_j\big) \\
&\quad + \sigma(a_1 - a_2 - \nu_j)\sin(2k\theta_j), && k = 0,\dots,m,\\
y_{2k+1}^{2j} &= 4b\sigma\cos(\theta_j)\sin\big((2k+1)\theta_j\big) \\
&\quad + b(a_2 - a_1 - \nu_j)\sin\big(2(k+1)\theta_j\big), && k = 0,\dots,m-1,
\end{aligned}$$

are eigenvectors for λ_{2j-1} *and* λ_{2j} *respectively. In addition , if* $\gamma = \frac{\sigma}{b^2}(2\sigma + a_2 - a_1) - 1$, *then an eigenvector for* $\lambda_{n+2} = a_2 + 2\sigma$ *is the vector* y^{n+2} *whose entries are*

$$\begin{aligned}
y_{2k}^{n+2} &= b^2U_k(\gamma) + \big(b^2 + \sigma(a_1 - a_2 - 2\sigma)\big)U_{k-1}(\gamma), && k = 0,\dots,m,\\
y_{2k+1}^{n+2} &= b\sigma U_{k-1}(\gamma), && k = 0,\dots,m-1.
\end{aligned}$$

Proof It is similar to the developed in the above propositions and corollaries. □

Remark 9 Observe that under the hypotheses of the above corollary, $\mathsf{J}(\mathsf{a},\mathsf{b})$ is a Jacobi 2-Toeplitz matrix iff $\sigma = 0$. In this case, $\theta_j = \frac{j\pi}{n+3}$

and $\gamma=-1$, which implies that $\big(1,0,-1,0,\dots,0,(-1)^m\big)$ is an eigenvector for $\lambda_{n+2}=a_2$. $j=1,\dots,m$. On the other hand, since $|a_2-a_1|<\nu_j$, $\lambda_j<\min\{a_1,a_2\}\le a_2\le\max\{a_1,a_2\}<\lambda_{m+1+j}$ for any $j=1,\dots,m$, and hence, a_1 is an eigenvalue of $\mathsf{J}(\mathsf{a},\mathsf{b})$ iff $a_2=a_1$. In this case, if $a=a_1=a_2$ again from the Identities (12) in the Appendix, we get that $ab^{n+1}U_m\Big(\frac{a^2}{2b^2}-1\Big)=b^{n+2}U_{n+2}\Big(\frac{a}{2b}\Big)$, valid even when $a=0$. Then, we recover the results obtained in Corollary 1.

We now analyze the remaining case, when either $\sigma=\sigma_+$ and $a_1>a_2+\frac{2b}{\sqrt{m(m+1)}}$ or $\sigma=\sigma_-$ and $a_2>a_1+\frac{2b}{\sqrt{m(m+1)}}$. Then for any $j=1,\dots,m$, we newly consider $\theta_j=\frac{1}{2}\theta_{m,j}(b^{-2}\sigma^2)$ given in Part 2 of Proposition 4 and also the positive real values $\nu_j=\sqrt{(a_2-a_1)^2+16b^2\cos^2\big(\theta_j\big)}$, $j=1,\dots,m-1$. The eigenvalues and the corresponding eigenvectors have the same expressions than in Corollary 5 except for λ_n and λ_{n+1} and also for y^n and y^{n+1}, that we describe in the next result.

Corollary 6 *Assume that either $\sigma=\sigma_+$ and $a_1>a_2+\frac{2b}{\sqrt{m(m+1)}}$ or $\sigma=\sigma_-$ and $a_2>a_1+\frac{2b}{\sqrt{m(m+1)}}$, and define define $\nu=\sqrt{(a_1-a_2)^2-16b^2\sinh^2\big(\theta_m\big)}$. Then, $\lambda_n=\frac{1}{2}(a_1+a_2-\nu)$ and $\lambda_{n+1}=\frac{1}{2}(a_1+a_2+\nu)$ are eigenvalues of $\mathsf{J}(\mathsf{a},\mathsf{b})$ and the vectors y^n and y^{n+1} whose entries are*

$$
\begin{aligned}
y^n_{2k} &= 4b^2\sinh(\theta_m)\cosh\big((2k+1)\theta_m\big)\\
&\quad -\sigma(a_1-a_2+\nu)\sinh(2k\theta_m), && k=0,\dots,m,\\
y^n_{2k+1} &= 4b\sigma\sinh(\theta_m)\cosh\big((2k+1)\theta_m\big)\\
&\quad +b(a_2-a_1+\nu)\sinh\big(2(k+1)\theta_m\big), && k=0,\dots,m-1,\\
y^{n+1}_{2k} &= 4b^2\sinh(\theta_m)\cosh\big((2k+1)\theta_m\big)\\
&\quad +\sigma(a_1-a_2-\nu)\sinh(2k\theta_j), && k=0,\dots,m,\\
y^{n+1}_{2k+1} &= 4b\sigma\sinh(\theta_m)\cosh\big((2k+1)\theta_m\big)\\
&\quad +b(a_2-a_1-\nu)\sinh\big(2(k+1)\theta_m\big), && k=0,\dots,m-1.
\end{aligned}
$$

6 Appendix: Chebyshev Polynomials

In this section we consider the definition and main properties about Chebyshev polynomials that have been used throughout the paper. We refer the reader to the book [25] for the explanation of the majority of the exposed results and moreover we include the prove of that properties that are lest usual and are not included in [25].

Any sequence of polynomials $\{P_k(x)\}_{k\in\mathbb{Z}}\subset\mathbb{R}[x]$ satisfying the *three-term recurrence*

$$-P_{k-1}(x)+2xP_k(x)-P_{k+1}(x)=0,\quad k\in\mathbb{Z},\tag{11}$$

is called *Chebyshev sequence (of polynomials)*. The first easy consequences of the three-term recurrence are

1. Any Chebyshev sequence $\{P_k(x)\}_{k\in\mathbb{Z}}$ is completely determined by the choice of the polynomials $P_0(x)$ and $P_1(x)$.
2. Given $\{P_k(x)\}_{k\in\mathbb{Z}}$ and $\{Q_k(x)\}_{k\in\mathbb{Z}}$ two Chebyshev sequences, then for any polynomials $r(x)$ and $s(x)$, $\{r(x)P_k(x)+s(x)Q_k(x)\}_{k\in\mathbb{Z}}$ is again a Chebyshev sequence. In particular, linear combinations of Chebyshev sequences are also Chebyshev sequences.

Observe that for any Chebyshev sequence of polynomials $\{P_k(x)\}_{k\in\mathbb{Z}}$, fixed $q\in\mathbb{R}$, $\{P_k(q)\}_{k\in\mathbb{Z}}$ is a solution of the Identity (8), the Chebyshev difference equation with coefficient q. This simple remark is of transcendental importance throughout the entire development of this paper.

Among the Chebyshev sequences, that named as *second kind* and denoted by $\{U_k(x)\}_{k\in\mathbb{Z}}$, plays a fundamental role. It can be defined as the unique Chebyshev sequence such that $U_0(x)=1$ and $U_1(x)=2x$. From the three-term recurrence we obtain the following properties for the second kind Chebyshev polynomials:

4. For any $k\in\mathbb{N}$, $U_k(x)$ has degree k and $U_{-1}(x)=0$.
5. $U_k(-x)=(-1)^kU_k(x)$ and $U_{-k}(x)=-U_{k-2}(x)$ for any $k\in\mathbb{Z}$.
6. $U_k(1)=k+1$, $U_{2k}(0)=(-1)^k$ and $U_{2k+1}(0)=0$, for any $k\in\mathbb{Z}$.

Another less obvious consequences of three-term recurrence are the identities

$$U_{2k+1}(x)=2xU_k(2x^2-1) \text{ and } U_{2k}(x)=U_k(2x^2-1)+U_{k-1}(2x^2-1), \quad (12)$$

valid for any $k\in\mathbb{Z}$, see the formulas (1.14) and (1.15) in [25].

The values of second order Chebyshev polynomials at the points of the interval $[-1,1]$ play a fundamental role. Since for $x\in[-1,1]$ there exists a unique $\theta\in[0,\pi]$ such that $x=\cos(\theta)$, we must evaluate U_m at $\cos(\theta)$. As a new consequence of the three-term recurrence, see [25, Sect 1.2.2], we have that

$$U_m\big(\cos(\theta)\big)=\frac{\sin\big((m+1)\theta\big)}{\sin(\theta)}, \text{ for any } m\in\mathbb{Z} \text{ and any } \theta\in(0,\pi). \quad (13)$$

Analogously, when $x>1$, there exists a unique $\Phi\in(0,+\infty)$ such that $x=\cosh(\Phi)$. Moreover, from the three-term recurrence we have that

$$U_m\big(\cosh(\Phi)\big)=\frac{\sinh\big((m+1)\Phi\big)}{\sinh(\Phi)}, \quad \text{for any} m\in\mathbb{Z} \text{and any} \Phi\in(0,+\infty). \quad (14)$$

L'Hôpital rule implies that the above identities are also true for $\theta=0,\pi$ and $\Phi=0$.

We end this section with some results about the zeroes of some Chebyshev polynomials obtained from the second kind ones. Specifically, for any $m\in\mathbb{N}$ and any $\beta\in\mathbb{R}$ we define the Chebyshev polynomial, whose degree is equal to m,

$$U_{m,\beta}(x)=U_m(x)+\beta U_{m-1}(x). \quad (15)$$

Whereas $U_{m,0}(x) = U_m(x)$, the polynomials $U_{m,-1}(x)$ and $U_{m,1}(x)$ are known respectively as *the third and fourth kind Chebyshev polynomials* of degree m.

From (13), the roots of $U_{m,0}(x) = U_m(x)$ are $z_{m,j} = \cos\big(\frac{j\pi}{m+1}\big)$, $j = 1,\dots,m$, and hence, all of them real and simple and moreover $1 > z_{m,1} > \cdots > z_{m,m} > -1$ since $\cos(\theta)$ is strictly decreasing in $[0,\pi]$. In addition, as $U_m(1) = m+1 > 0$, $U_m(x) > 0$ when $x > z_{m,1}$ and in particular $\lim\limits_{x\to+\infty} U_m(x) = +\infty$. Analogously, as $U_m(-1) = (-1)^m$, when $x < z_{m,m}$, then $(-1)^m U_m(x) > 0$ and hence $\lim\limits_{x\to-\infty} (-1)^m U_m(x) = +\infty$. Therefore, for any $\beta \in \mathbb{R}$, we have that $\lim\limits_{x\to+\infty} U_{m,\beta}(x) = \lim\limits_{x\to-\infty} (-1)^m U_{m,\beta}(x) = +\infty$.

Proposition 4 *Given $m \geq 1$ and $\beta \in \mathbb{R}$ the Chebyshev polynomial $U_{m,\beta}(x)$ has m real and simple roots $z_{m,1}(\beta) > \cdots > z_{m,m}(\beta)$ and the following properties hold:*

1. *If $\beta = 0$, then $z_{m,j}(0) = \cos\big(\theta_j(0)\big)$ where $\theta_{m,j}(0) = \frac{j\pi}{m+1}$, for any $j = 1,\dots,m$.*
2. *If $\beta > 0$, for any $j = 1,\dots,m-1$ there exist $\frac{j\pi}{m+1} < \theta_{m,j}(\beta) < \frac{(j+1)\pi}{m+1}$ such that $z_{m,j}(\beta) = \cos\big(\theta_j(\beta)\big)$. Moreover, there exists $\theta_{m,m}(\beta) > 0$ satisfying that:*
   (a) When $0 < \beta < 1 + \frac{1}{m}$, $\frac{m\pi}{m+1} < \theta_{m,m}(\beta) < \pi$ and $z_{m,m}(\beta) = \cos\big(\theta_{m,m}(\beta)\big)$.
   (b) When $\beta = 1 + \frac{1}{m}$, $\theta_{m,m}(\beta) = \pi$ and $z_{m,m}(\beta) = \cos\big(\theta_{m,m}(\beta)\big) = -1$.
   (c) When $\beta > 1 + \frac{1}{m}$, $\theta_{m,m}(\beta) < \log(\beta)$ and $z_{m,m}(\beta) = -\cosh\big(\theta_{m,m}(\beta)\big)$.
3. *If $\beta < 0$, for any $j = 2,\dots,m$ there exist $\frac{(j-1)\pi}{m+1} < \theta_{m,j}(\beta) < \frac{j\pi}{m+1}$ such that $z_{m,j}(\beta) = \cos\big(\theta_j(\beta)\big)$. Moreover, there exists $\theta_{m,1}(\beta) \geq 0$ satisfying that:*
   (a) When $-1 - \frac{1}{m} < \beta < 0$, $0 < \theta_{m,1}(\beta) < \frac{\pi}{m+1}$ and $z_{m,1}(\beta) = \cos\big(\theta_{m,1}(\beta)\big)$.
   (b) When $\beta = -1 - \frac{1}{m}$, $\theta_{m,1}(\beta) = 0$ and $z_{m,1}(\beta) = \cos\big(\theta_{m,1}(\beta)\big) = 1$.
   (c) When $\beta < -1 - \frac{1}{m}$, $0 < \theta_{m,1}(\beta) < \log(|\beta|)$ and $z_{m,1}(\beta) = \cosh\big(\theta_{m,1}(\beta)\big)$.

Proof Clearly $z_{m,j}(0) = z_{m,j}$, $j = 1,\dots,m$, and hence the result for $\beta = 0$ follows. On the other hand, since $U_{m,-\beta}(x) = (-1)^m U_{m,\beta}(-x)$ from property (5), then z is a root of $U_{m,-\beta}(x)$ iff $-z$ is a root of $U_{m,\beta}(x)$. Therefore part 3 is consequence of part 2, so we only proof this case.

From identity (13), for any $j = 1,\dots,m$, we have that $U_{m-1}(z_{m,j}) = (-1)^{j+1}$. Therefore, if $j = 1,\dots,m-1$, then $U_{m,\beta}(z_{m,j})U_{m,\beta}(z_{m,j+1}) = -\beta^2$, which implies that $U_{m,\beta}(x)$ has a root $z_{m,j}(\beta)$ in the interval $(z_{m,j+1}, z_{m,j})$. This identifies $m-1$ roots of $U_{m,\beta}(x)$ as real and simple, so the remaining root $z_{m,m}$ must be real.

Because $U_{m,\beta}(z_{m,m})U_{m,\beta}(-1) = -\beta(m+1-\beta m) < 0$ iff $0 < \beta < 1 + \frac{1}{m}$, if the above condition holds, then $z_{m,m} \in (z_{m,m}, -1)$. On the other hand, since $(-1)^m U_{m,\beta}(-1) = m + 1 - \beta m$, it follows that $z_{m,m}(\beta) = -1$ when $\beta = 1 + \frac{1}{m}$ and also that $(-1)^m U_{m,\beta}(-1) < 0$ when $\beta > 1 + \frac{1}{m}$. Besides, from Identity (14) we have that $(-1)^m U_{m,\beta}\big(-\cosh(\log(\beta))\big) = \frac{1}{\beta^m} > 0$ and hence we conclude that $-\cosh\big(\log(\beta)\big) < z_{m,m}(\beta) < -1$.

The existence of the claimed $\theta_{m,j}(\beta)$ is consequence of beeing $\cos(\theta)$ strictly decreasing in $[0, \pi]$ and also of being $\cosh(x)$ strictly increasing from $(0, +\infty)$ onto $(1, +\infty)$. □

It is well known that , for any $j = 1, \ldots, m$, we have that $\theta_{m,j}(1) = \frac{2j\pi}{2m+1}$ and hence, $z_{m,j}(1) = \cos\left(\frac{2j\pi}{2m+1}\right)$, see [25, Identity (2.8)], whereas $\theta_{m,j}(1) = \frac{(2j-1)\pi}{2m+1}$ and hence, $z_{m,j}(-1) = \cos\left(\frac{(2j-1)\pi}{2m+1}\right)$, see [25, Identity (2.7)].

Acknowledgements This work has been partly supported by the Spanish Research Council (Comisión Interministerial de Ciencia y Tecnología,) under project PID2021-122501NB-I00 and by the Universitat Politècnica de Catalunya under funds AGRUPS-2023 and AGRUPS-2024. Samir Mondal acknowledges funding received from the Prime Minister's Research Fellowship (PMRF), Ministry of Education, Government of India, for carrying out this work.

References

1. Álvarez-Nodarse, R., Petronilho, J., Quintero, N.R.: Spectral properties of certain tridiagonal matrices. Linear Algebra Appl. **436**, 682–698 (2012)
2. Avila, A., Jitomirskaya, S.: Solving the ten martini problem. In: Mathematical Physics of Quantum Mechanics. Lecture Notes in Physics, vol. 690, pp. 5–16. Springer, Berlin (2006)
3. Avila, A., Jitomirskaya, S.: The ten martini problem. Ann. Math. **170**(2), 303–342 (2009)
4. Avni, N., Breuer, J., Kalai, G., Simon, B.: Periodic boundary conditions for periodic Jacobi matrices on trees. Pure Appl. Funct. Anal. **7**, 489–502 (2022)
5. Bendito, E., Encinas, A.M., Carmona, A.: Eigenvalues, eigenfunctions and Green's functions on a path via Chebyshev polynomials. Appl. Anal. Discret. Math. **3**, 282–302 (2009)
6. de Boor, C., Golub, G.H.: The numerically stable reconstruction of a Jacobi matrix from spectral data. Linear Algebra Appl. **21**, 245–260 (1978)
7. Chang, H.-W., Liu, S.-E., Burridge, R.: Exact eigensystems for some matrices arising from discretizations. Linear Algebra Appl. **430**, 999–1006 (2009)
8. Cuminato, J.A., McKee, S.A.: A note on the eigenvalues of a special class of matrices. J. Comput. Appl. Math. **234**, 2724–2731 (2010)
9. Encinas, A.M., Jiménez, M.J.: Floquet Theory for second order linear difference equations. J. Diff. Eq. App. **22**, 353–375 (2016)
10. Encinas, A.M., Jiménez, M.J.: Second order linear difference equations. J. Differ. Equ. Appl. **24**, 305–343 (2018)
11. Encinas, A.M., Jiménez, M.J.: Bounded solutions of self-adjoint second order linear difference equations with periodic coefficients. Open Math. **16**, 75–82 (2018)
12. Encinas, A.M., Jiménez, M.J., Marijuán, C., Mitjana, M., Pisonero, M. Bisymmetric nonnegative Jacobi matrix realizations. Linear Multilinear Algebra, published on-line, pp. 1–28. https://doi.org/10.1080/03081087.2023.2297391
13. da Fonseca, C.M., Petronilho, J.: Explicit inverses of some tridiagonal matrices. Linear Algebra Appl. **325**, 7–21 (2001)
14. da Fonseca, C.M., Petronilho, J.: Explicit inverse of a tridiagonal k-Toeplitz matrix. Numer. Math. **100**, 457–482 (2005)
15. Gladwell, G.M.L.: Inverse Problems in Vibration. Solid Mechanics and its Applications, vol. 119. Kluwer Academic Publishers, Dordrecht (2004)
16. Gladwell, G.M.L., Jones, T.H., Willms, N.B.: A test matrix for an inverse eigenvalue problem. J. Appl. Math. 515082, 6 (2014)

17. Gray, L.J., Wilson, D.G.: Construction of a Jacobi Matrix from spectral data. Linear Algebra Appl. **14**, 131–134 (1976)
18. Gover, M.J.C.: The eigenproblem of a tridiagonal 2-Toeplitz Matrix. Linear Algebra Appl. **197**(198), 63–78 (1994)
19. Gregory, R.T., Karney, D.L.: A collection of matrices for testing computational algorithms. Robert E. Krieger Publishing Co., Huntington, N.Y. (1978)
20. Hald, O.H.: Inverse eigenvalue problems for Jacobi Matrices. Linear Algebra Appl. **14**, 63–85 (1976)
21. Hochstadt, H.: On the construction of a Jacobi Matrix from spectral data. Linear Algebra Appl. **8**, 435–446 (1974)
22. Kiliç, E., Arikan, T.: Evaluation of spectrum of 2-periodic tridiagonal-Sylvester matrix. Turkish J. Math. **40**, 80–89 (2016)
23. Last, Y.: Spectral theory of Sturm-Liouville operators on infinite intervals: a review of recent developments. In: Amrein, O., Hinz, A.M., Pearson, D.B. (eds.) Sturm-Liouville Theory: Past and Present W, pp. 99–120. Basel, Birkhäuser (2005)
24. Marcellán, F., Petronilho, J.: Eigenproblems for tridiagonal 2-Toeplitz matrices and quadratic polynomial mappings. Linear Algebra Appl. **260**, 169–208 (1997)
25. Mason, J.C., Handscomb, D.: Chebyshev polynomials. Chapman & Hall, CRC Press Company, Boca de Ratón (2003)
26. Meyer, C.D.: Matrix analysis and applied linear algebra. In: Society for Industrial and Applied Mathematics (SIAM), Philadelphia, PA (2000)
27. van Mouche, P.: The coexistence problem for the discrete Mathieu operator. Commun. Math. Phys. **122**, 23–33 (1989)
28. Puig, J.: Cantor spectrum for the almost Mathieu operator. Commun. Math. Phys. **244**, 297–309 (2004)
29. Simon, B.: Schrödinger operators in the twenty-first century. In: Fokas, A., Grigoryan, A., Kibble, T., Zegarlinski, B. (eds.) Mathematical Physics, pp. 283–288. Imperial College Press, London (2000)
30. Willms, A.R.: Analitic results for the eigenvalues of certain tridiagonal matrices. SIAM J. Matrix Anal. Appl. **30**, 639–656 (2008)
31. Yueh, W.-C.: Eigenvalues of several tridiagonal matrices. Appl. Math. E-Notes **5**, 66–74 (2005)

Convolutional Neural Networks for Anomaly Classification and Localization Using Neutron Noise

E. Navarro-Gamón, A. Vidal-Ferràndiz, M. Chillarón, and G. Verdú

Abstract Classifying and localizing anomalies in the performance of nuclear reactors is an issue to assure the safety of nuclear reactors and, consequently, guarantee their long-term operations. It is of particular interest to detect perturbations while are occurring without disturbing the operation of the nuclear reactor. This detection can be done with the analysis of neutron flux oscillations around a steady state value, called neutron noise. In this work. we propose a deep learning framework for the deconvolution of reactor transfer functions from perturbation-induced neutron noise sources. The main objective of this work is to develop tools based on deep learning techniques using convolutional neural networks, working with simulated data with different Gaussian noise levels, and to study the number and location of detectors that need to be active. In particular, the data used has been simulated using the neutron diffusion approximation and the first-order noise approximation for the BIBLIS 2D reactor and IAEA 3D reactor. High-accuracy results are obtained both when predicting the type of the perturbation and when locating the place of the perturbation, with a low error rate even when only four to eight detectors are available.

Keywords Neutron noise · Deep learning · Convolutional neural networks · Nuclear reactor monitoring · Anomaly detection · Perturbation localization

1 Introduction

Neutron noise analysis examines the impact of minor disturbances in nuclear reactors by utilizing the natural fluctuations in neutron flux captured by core instruments, referred to as *neutron noise*. These disturbances typically manifest as variations in macroscopic cross-sections and may arise from random changes in coolant properties and the mechanical vibrations of core components [16]. This analysis is employed for non-invasive monitoring and the identification and detection of anomalies in

E. Navarro-Gamón (✉) · A. Vidal-Ferràndiz · M. Chillarón · G. Verdú
Universitat Politècnica de València, València, Spain
e-mail: emnaga@upvnet.upv.es

J. R. Torregrosa (ed.), *Mathematical Modelling in Engineering Human Behaviour*, Springer Proceedings in Mathematics & Statistics 513,
https://doi.org/10.1007/978-3-032-00567-0_2

nuclear power plants. Identifying anomalies in nuclear reactors is crucial to ensure the safety of the facilities and prevent mishaps. In recent years, numerous methods have been formulated to simulate and recognize anomalies using neutron noise in nuclear reactors. Particularly, the integration of deep learning techniques has shown to be successful for the prompt detection of such anomalies.

Multiple studies have demonstrated encouraging outcomes. For example, Tasakos et al. [18] propose a deep learning approach to study such anomalies. Twelve or forty-eight perturbation location sources were initially extracted from the signal using a CNN. Subsequently, the resolution of the unfolding was significantly enhanced via a k-means clustering and k-Nearest Neighbor coarse-to-fine approach. Additionally, reactor power signals exposed to various noise and corruption levels were denoised and reconstructed with a DAE (denoising autoencoder). The reconstructed signals were evaluated against the original ones using normalized cross-correlation and unfolding measures. The findings indicate that, even with limited training data and obscured or noisy signals, the perturbation sources can be localized with remarkable accuracy at multiple granularity levels.

In their work, Kollias et al. [10] present a novel approach utilizing machine learning for noise diagnostics in nuclear plant reactor cores, relying solely on neutron detector measurements. They utilize advanced core simulation tools based on diffusion principles, particularly the CORE SIM+ modelling framework, to produce comprehensive simulated datasets with various perturbations originating from distinct core areas [13]. By training cutting-edge machine learning and deep learning models on this data, they successfully accomplish semantic segmentation, classification, and localization of multiple simultaneous core perturbations. This technique is then tested on real plant data from two reactors that lack perturbation labels. To bridge the gap between simulated and real plant data, a domain adaptation methodology is created, employing self-supervised or unsupervised learning strategies to align the datasets and identify, classify, and locate perturbations. Experimental outcomes highlight the efficacy of the proposed approach in utilizing simulated data to predict real plant environments. Likewise, Calivá et al. [2] employ a mix of denoising autoencoders (DAE), k-means clustering, and CNNs. The findings indicate that the sources of disturbances can be pinpointed with high precision, even with limited training data and under noisy or difficult detection conditions.

Papaoikonomou et al. [14] propose a deep learning approach to operate in the setting of unsupervised real plant measurements, where information about the true perturbation characteristics is unknown. A Swiss 3-loop pre-KONVOI reactor is simulated to train neural networks under various perturbation conditions, illustrating the application of a self-supervised domain adaptation method to accurately align the neural network's representations between synthetic and actual detector data for better identification and localization of perturbations. This self-supervised method identifies shared features in both synthetic and real datasets through structural and non-semantic tasks, eliminating the need for human annotations. It is demonstrated that machine learning algorithms can derive reactor transfer functions from time-series neutron detector measurements in both the simulated pre-training phase and real-world training. The findings also suggest that algorithms pre-trained with simulated data can effectively predict real plant scenarios employing self-supervised domain adaptation techniques. This research is part of the CORTEX project, standing for Core Monitoring Techniques and Experimental validation and demonstration, aimed at developing an innovative core monitoring approach for detecting anomalies in nuclear reactors, such as excessive vibrations of core internals, flow blockages, and coolant inlet perturbations.

Other works like Kamkar and Abasi [9] study different types of machine learning approaches from spanning decision trees, random forest, k-nearest neighbours, multilayer perceptron, support vector machine, and 1D-convolutional neural network. This work concludes that the multilayer perceptron, the support vector machine and the 1D-convolutional neural network show the best accuracy results.

In other way, Ejigu and Liu [5] monitor the PWR core power and outlet temperature using a gradient descent-particle swarm optimization, a hybrid algorithm-based deep neural network (GD-PSO-based DNN) technique. Incorporating Gaussian noise into the input signal, a Lyapunov analysis is used to look at the PWR core stability criteria. The simulation results show that the suggested control strategy outperforms the sliding mode control, linear quadratic regulator, and proportional-integral-derivative methods in terms of tracking reference inputs successfully and improve stability. Data from RELAP5 system codes are used to validate the controller and many error metrics are used to assess performance. The benefit of using the GD-PSO-based DNN control method for system identification, prediction, and control in various nuclear engineering disciplines is provided by this work.

All in all, the research interest in deep learning applied to the study and monitoring of nuclear power plants is growing with all the promising studies. In this work, we propose and study two deep learning models to locate and classify neutron noise anomalies when a minimal number of neutron detectors is available. This model is based in the VGG16 neural network [2, 17] a multiclass image classification. In the case of our work, the parameters and hyperparameters are updated according to the necessities of our data, and the number of hidden layers is selected experimentally to reduce computational time but maintaining the capability to extract discriminative features.

The document is organized as follows: In Sect. 2, the simulation of neutron noise data is explained. The data is then adapted for their use with neural networks, as explained in Sect. 3.1. Section 3.2 presents the neural network designs achieved for signal classification, the results of which are shown in Sect. 4. Finally, Sect. 5 presents the conclusions of the developed work.

2 Neutron Noise Simulation of Anomalies

The data to train the neural networks is generated using FEMFFUSION code. This C++ open-source library is developed by Universitat Politècnica de València [6, 20].

It uses the finite element method to solve the diffusion approximation of the neutron transport equation for commercial and research reactors, from any generation, even Small Modular Reactors (SMR), and for all types of geometries and dimensions [3, 21].

FEMFFUSION simulates the perturbation solving the first-order neutron noise steady state equation in the frequency domain and using the diffusion approximation, which can be written as in Eq. (1) [22].

$$\mathcal{A}\delta\Phi = \mathcal{B}\Phi_0, \tag{1}$$

This partial differential equation requires the steady state solution of the neutron flux ϕ_0 as well as its associated weights and corresponding static eigenvalue. Then, taking $\mathcal{A}$ and $\mathcal{B}$ as in Eqs. (2) and (3), the neutron noise flux $\delta\phi$ can be obtained [22].

$$\mathcal{A} = \begin{bmatrix} \frac{i\omega}{v_1} - \nabla D_1 \nabla + \Sigma_{a1}^0 + \Sigma_{12}^0 - \gamma\nu\Sigma_{f1}^0 & -\gamma\nu\Sigma_{f2}^0 \\ -\Sigma_{12}^0 & \frac{i\omega}{v_2} - \nabla D_2 \nabla + \Sigma_{a2} \end{bmatrix}, \tag{2}$$

$$\mathcal{B} = \begin{bmatrix} -\delta\Sigma_{a1} - \delta\Sigma_{12} + \gamma\delta\nu\Sigma_{f1} & +\gamma\delta\nu\Sigma_{f2} \\ \delta\Sigma_{12} & -\delta\Sigma_{a2} \end{bmatrix}. \tag{3}$$

From the simulations in FEMFFUSION, a file for each type of perturbation, assembly and frequency is generated. The files contain the information of the neutron power and the neutron flux for the first group energy or fast flux (ϕ_1), the neutron noise flux for the first group energy ($\delta\phi_1$), the neutron flux for the second group of energy or thermal flux (ϕ_2) and the neutron noise flux for the second group of energy ($\delta\phi_2$). However, only ϕ_2 and $\delta\phi_2$ are the simulated values that are considered to train the convolutional networks, particularly the evaluation in Eq. (4) which depicts the transfer function of the detectors for the thermal neutron flux inside the core.

$$s = |\delta\phi_2|/\phi_2 \tag{4}$$

An example of the static thermal flux ϕ_2 and the neutron noise $|\delta\phi_2|/\phi_2$ distributions obtained from the FEMFFUSION files is presented in Fig. 1, for the assembly (8, 4) when the scattering cross-section value has been changed.

3 Convolutional Neural Networks

3.1 Data Preprocessing

In deep learning techniques based on the convolutional neural networks used in this work, the dataset simulated to train the models needs to be broad and must represent the matter in hand. Then, several data augmentation techniques [7, 19] are proposed to assure that the networks have enough data to avoid overfitting and improve the

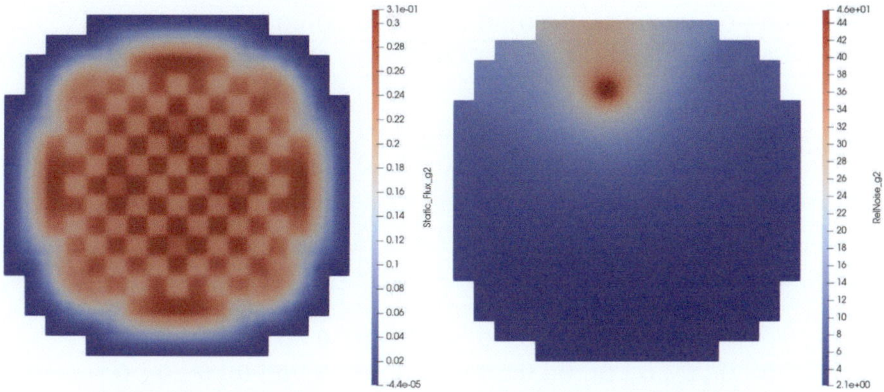

(a) Static flux ϕ_2 (A.U.) (b) Neutron noise $|\delta\phi_2|/\phi_2$ (%)

Fig. 1 FEMFFUSION results for a change of the scattering cross-section in assembly at position (8, 4) on the BIBLIS reactor

quality of the training dataset. On one hand, several distributions of active detectors are applied to study the minimum number of active detectors that are needed to obtain high accuracies. On the other hand, Additive White Gaussian Noise (AWGN) is added to the signals to introduce the possible errors in the measurements of the detectors and increase the training dataset.

For the distribution of the detectors, a total of 12 different sets of arrays of active detectors, called masks, are applied. Each detector is located in one of the assemblies and its value is considered when the detector in that position is set to the unity. The active detectors are evenly distributed and distributions of 4, 8, 16, 24, 32, 46, 64, 82, 97, 128, 184 and all assemblies active are applied. An example of the mask of 8 active detectors is shown in Fig. 2. Despite the application of the masks of active detectors, the input data for the fitting models is still distributed in arrays of 17×17 elements, being nil in the position where the detector is not activated or if no assembly is present in it.

The application of the masks of active detectors generates a new dataset to which the AWGN is added. This noise is added to the signal s of Eq. (6) in percentage of noise level, starting from 5 to 25% in increments of 2.5%. With this technique, two sets of signals can be used to train the convolutional neural network: the clean signals and the noisy signals. Figures 3 and 4 compare, for the 2D case of study, the clean and noisy signals for the scattering and absorption cross-sections perturbations, located in the 140th assembly and for a frequency of oscillation of 1.0 Hz.

Similarly, the addition of 25% of white Gaussian noise for the three-dimensional reactor is compared with the clean signals in Figs. 5, 6 and 7. In Fig. 5 is represented the axial perturbation in the 140th element when the velocity of the coolant is changed to 300 m/s and oscillates with a frequency of 1.0 Hz. For the scattering perturbation, Fig. 6 represents the comparison between the clean and noisy signals when the

0	0	0	0	0	0	0	0	0	0	0	0	0	0	0	0	0
0	0	0	0	0	0	0	0	0	0	0	0	0	0	0	0	0
0	0	0	0	0	0	0	0	0	0	0	0	0	0	0	0	0
0	0	0	0	0	0	0	0	0	0	0	0	0	0	0	0	0
0	0	1	0	0	0	0	0	0	0	0	0	0	0	1	0	0
0	0	0	0	0	0	0	0	1	0	0	0	0	0	0	0	0
0	0	0	0	0	0	0	0	0	0	0	0	0	0	0	0	0
0	0	0	0	0	0	0	0	0	0	0	0	0	0	0	0	0
0	0	0	0	0	1	0	0	0	0	0	1	0	0	0	0	0
0	0	0	0	0	0	0	0	0	0	0	0	0	0	0	0	0
0	0	0	0	0	0	0	0	0	0	0	0	0	0	0	0	0
0	0	0	0	0	0	0	0	1	0	0	0	0	0	0	0	0
0	0	1	0	0	0	0	0	0	0	0	0	0	0	1	0	0
0	0	0	0	0	0	0	0	0	0	0	0	0	0	0	0	0
0	0	0	0	0	0	0	0	0	0	0	0	0	0	0	0	0
0	0	0	0	0	0	0	0	0	0	0	0	0	0	0	0	0
0	0	0	0	0	0	0	0	0	0	0	0	0	0	0	0	0

Fig. 2 Distribution of the 8 active detectors in the 2D mask of active detectors

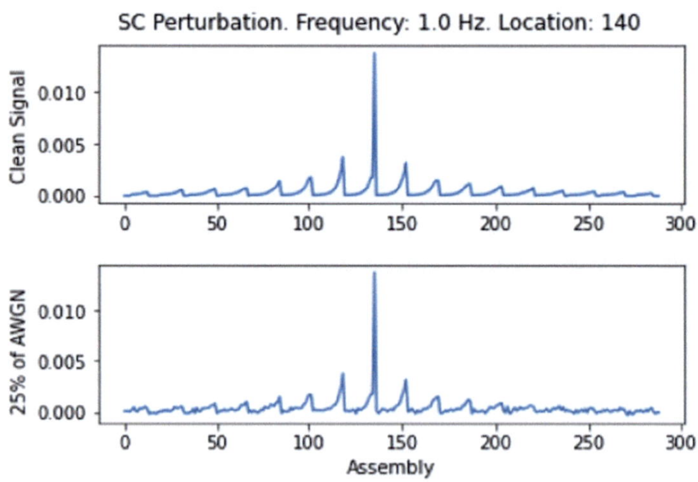

Fig. 3 Clean signal and noisy signal comparison for the *Scattering Cross-section Perturbation*. 2-dimensional reactor

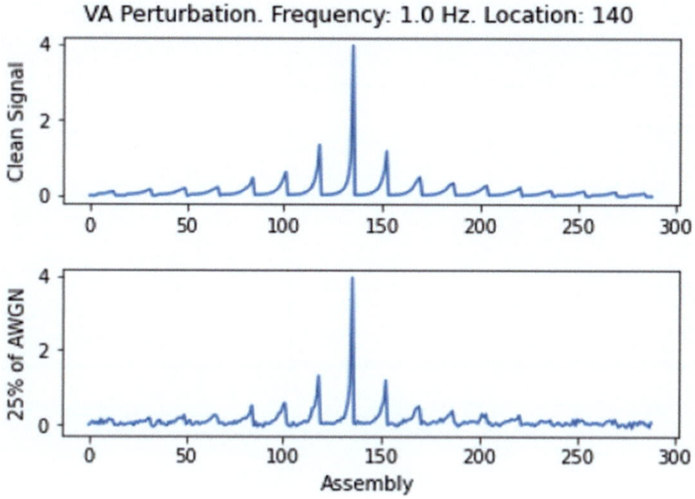

Fig. 4 Clean signal and noisy signal comparison for the *Absorption Cross-section Perturbation*. 2-dimensional reactor

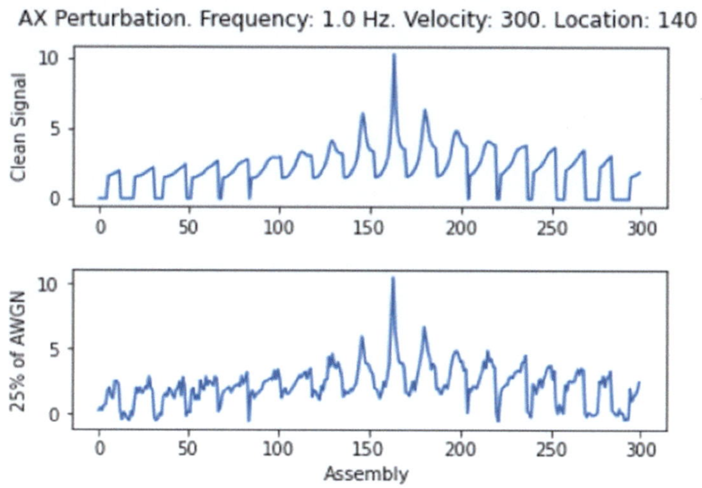

Fig. 5 Clean signal and noisy signal comparison for the *Axial Perturbation*. 3-dimensional reactor

cross-section oscillates with a frequency of 1.0 Hz in the location 1350, and Fig. 7 represents the comparison for the absorption cross-section perturbation.

The addition of the noise level involves several steps that start with the calculation of the power of the initial signal P_s as the mean squared value as in Eq. (5).

$$P_s = \frac{1}{n} \sum_{k=1}^{n} s(k)^2 \tag{5}$$

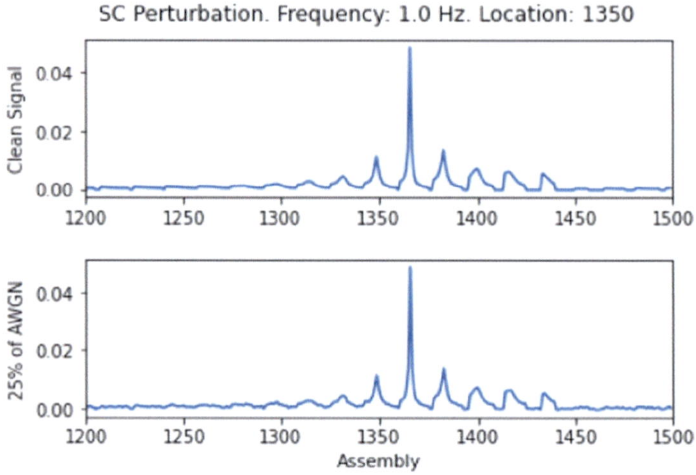

Fig. 6 Clean signal and noisy signal comparison for the *Scattering Cross-section Perturbation*. 3-dimensional reactor

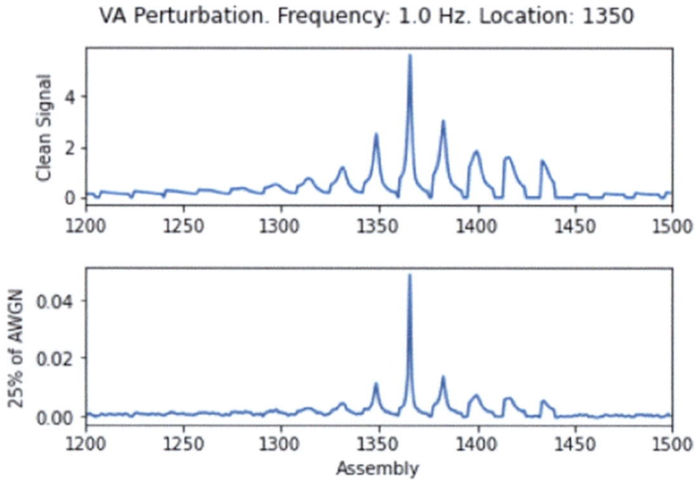

Fig. 7 Clean signal and noisy signal comparison for the *Absorption Cross-section Perturbation*. 3-dimensional reactor

Then the noise level l is added in percentage in levels that go from 5 to 25% in increments of 2.5%, obtaining the noise power P_n as in Eq. (6).

$$P_n = l \cdot P_s \tag{6}$$

To introduce the Gaussian noise, a normal distribution of mean 1 and standard deviation of zero ($\varphi =\sim \mathcal{N}(0, 1)$) is generated and used to adjust the noise power as in Eq. (7).

$$\bar{\varphi} = \sqrt{\frac{P_n}{\frac{1}{n}\sum_{k=1}^{n}\varphi(k)^2}} \cdot \varphi \tag{7}$$

This adjusted noise power $\bar{\varphi}$ is scaled with the corresponding percentage of noise level l as in Eq. (8).

$$\varphi_l = l \cdot \bar{\varphi} \tag{8}$$

In Eq. (9), the scaled noise power is added to the original signal s, finally obtaining the noisy signal s_φ.

$$s_\varphi = s + \varphi_l \tag{9}$$

3.2 Design of the Convolutional Neural Networks Models

With the data correctly prepared, the inputs to train the networks are complete. However, the dataset must be split into a set for training, a set for validating, and another for testing the trained model. For the validation, 10% of the signals are kept, and their aim is to evaluate the trained model with an unbiased perspective while the hyperparameters are tuned. 15% of the input signals are for testing the validated model, using data that the model has not seen before. The tool used to proceed with the data splitting is the *train_test_split* from the Scikit-learn [15]. This library allows splitting the data while it is shuffled, and the percentage of each set of data is equally distributed when the stratified option is enabled.

The convolutional neural network models are based on the architecture of the multi-class image classification network proposed by Karen Simonyan and Andrew Zisserman in the VGG16 model [17]. However, it has been adapted to the cases of study treated in this work and to neutron noise problems. The number of convolutional layers is reduced and adapted to match the 2D and 3D convolutions and input dimensions, since no longer treats RGB images with 244 × 244 dimensions. Also, the filters are smaller, reducing the computational time and complexity without compromising the most important characteristics of the signals.

For classifying the type of perturbation in the 3D case, the chosen architecture is pictured in Fig. 8 whereas to locate the perturbation in the 3D reactor, the architecture is described in Fig. 9. The models are composed of 4 convolutional layers activated with ReLU activation functions and include Batch Normalization, which transforms the output data to have a mean close to 0 and standard deviation of nearly 1. Following the convolutional layers, the Max Pooling steps are applied to reduce the sampling. Then, the output shape is turned into a one-dimensional vector and introduced in 2 fully connected layers, activated with ReLU functions and Batch Normalization. However, to avoid overfitting and generalize the training, a Dropout step is included. The Dropout technique sets, randomly, part of the neurons to 0, so those do not participate during the training and oblige the neurons to recombine. Finally, the last layer is the activation function to classify the output data. Since only 2 types of

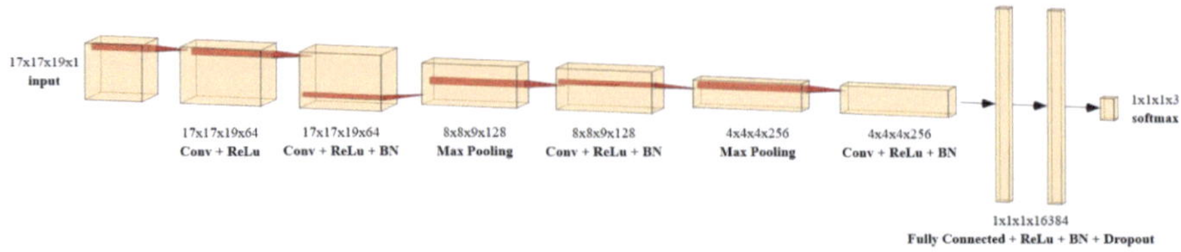

Fig. 8 Three-dimensional CNN architecture to classify the type of perturbation

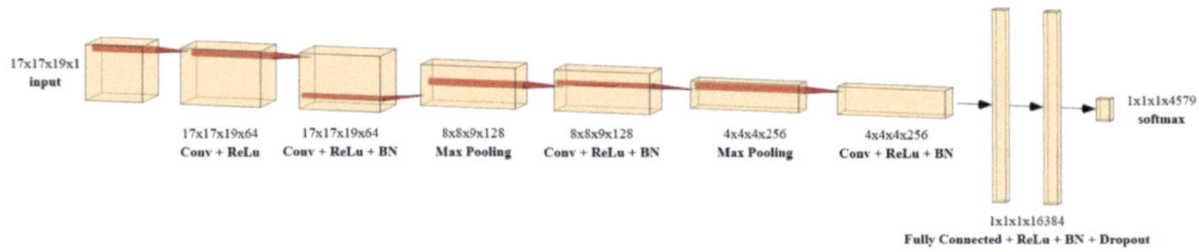

Fig. 9 Three-dimensional CNN architecture to locate the perturbation

perturbations are considered in the two-dimensional reactor, the activation function must be a Sigmoid, whilst the activation function must be a Softmax classification for locating the perturbations and identifying the type of perturbation in the three-dimensional case.

The filter dimensions can be seen in Figs. 8 and 9, the Dropout rate is set to 0.3 so 30% of the neurons are neglected and the learning rate is set to $1.0e^{-4}$ for identifying the type of perturbation and $5.0e^{-5}$ for locating the perturbations. The hyperparameters are set experimentally, the optimizer that is used is Adam, and the loss functions are binary cross-entropy for the classification of the perturbations in the 2D model and categorical cross-entropy for the rest of the trainings. Finally, the batch size is set to 32 and the early stopping technique is used to reduce computation time. The training library used is Keras [4] from TensorFlow.

4 Numerical Results

For studying the application of convolutional neural networks to locate and identify the type of perturbations in nuclear reactors, two cases of study are considered. First, the two-dimensional BIBLIS reactor is considered. Secondly, the three-dimensional IAEA reactor is studied. To both types of reactors, several perturbations are introduced.

The chosen perturbations to simulate the neutron noise are small variations of the absorption and scattering cross-sections, which oscillate at frequencies $f =$ 0.1, 0.5, 1, 5 and 10 Hz. The absorption cross-section perturbation is introduced as in Eq. (10), similarly to the scattering cross-section perturbation that is simulated as in Eq. (11).

$$\Sigma_{a,g}(t) = \Sigma_a(0) + 0.01\,\Sigma_a(0)\sin(2\pi f t) \tag{10}$$

$$\Sigma_{s,g}(t) = \Sigma_s(0) + 0.01\,\Sigma_s(0)\sin(2\pi f t) \tag{11}$$

Moreover, for the 3D case of study, an axial perturbation is considered. This perturbation is performed as a variation of the velocity of the coolant flow, making the neutron flux oscillate at the same frequencies and coolant velocities $v =$ 100, 200, 300, 400 and 500 m/s. Being all these different perturbations applied to every assembly, the dataset to train the convolutional neural network models results in 2570 neutron noise simulations for the two-dimensional case and 51815 simulations for the three-dimensional case of study.

4.1 Two-Dimensional BIBLIS Reactor

The BIBLIS 2D geometry is given in Fig. 10 [8, 11], which represents a reactor of 257 assemblies, arranged in a checkerboard shape. Two energy group cross sections can be found in Table 1. The reactor has eight homogenized materials, where the outermost represents the reflector. Vacuum boundary conditions are applied at the external boundary. Reference solution for the steady state are given in [12].

Fitting the convolutional neural network models takes using the training and validation sets of data. However, the test set of signals is not used for the training process. This unseen data is used to evaluate the accuracy that the model reaches after the training process.

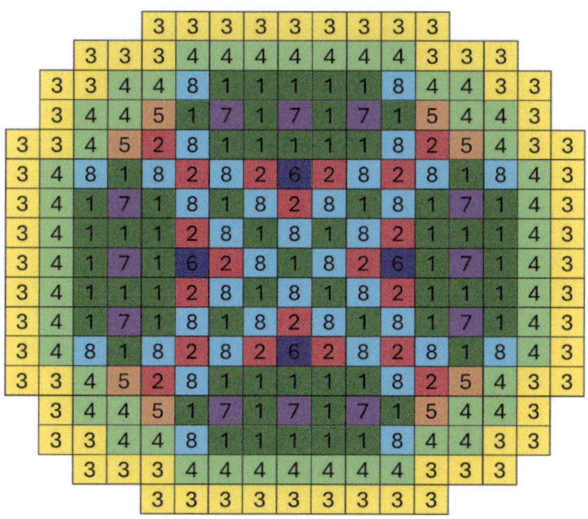

Fig. 10 Reactor geometry of the BIBLIS 2D benchmark

Table 1 Macroscopic cross sections of the BIBLIS 2D reactor

Material	g	D_g (cm)	Σ_{ag} (1/cm)	$\nu\Sigma_{fg}$ (1/cm)	Σ_{fg} (1/cm)	Σ_{12} (1/cm)
1	1	1.4360	0.0095042	0.0058708	0.0023768	0.017754
	2	0.3635	0.0750580	0.0960670	0.0388940	
2	1	1.4366	0.0096785	0.0061908	0.0025064	0.017621
	2	0.3636	0.0784360	0.1035800	0.0419350	
3	1	1.3200	0.0026562	0.0000000	0.0000000	0.023106
	2	0.2772	0.0715960	0.0000000	0.0000000	
4	1	1.4389	0.0103630	0.0074527	0.0030173	0.017101
	2	0.3638	0.0914080	0.1323600	0.0535870	
5	1	1.4381	0.0100030	0.0061908	0.0025064	0.017290
	2	0.3665	0.0848280	0.1035800	0.0419350	
6	1	1.4385	0.0101320	0.0064285	0.0026026	0.017192
	2	0.3665	0.0873140	0.1091100	0.0441740	
7	1	1.4389	0.0101650	0.0061908	0.0025064	0.017125
	2	0.3679	0.0880240	0.1035800	0.0419350	
8	1	1.4393	0.0102940	0.0064285	0.0026026	0.017027
	2	0.3680	0.0905100	0.1091100	0.0441740	

The results for the two-dimensional reactor geometry trainings are presented in Table 2. These trainings were performed applying the different masks of detectors and considering all the files generated with the FEMFFUSION code. Regarding the architecture of the neural networks, the layers distribution is the same as in Fig. 8 for the classification training and Fig. 9 for the location training, although the dimensions of the inputs are 2D arrays of 17×17 and the convolutions are performed for the same dimensions.

As Table 2 shows, both training networks present a tendency of increasing the accuracy when more detectors are active. This behavior is confirmed in Fig. 9, demonstrating that the classification training reaches the 100% accuracy faster than the location training, specifically with 16 active detectors. These results are expected, since a high number of active detectors allows having more data to proceed with the training, and the classification model only needs to choose between 2 perturbations instead of the 257 possible locations. With these network architectures, the results for the accuracy have proven to be high enough to characterize the perturbations in these two-dimensional cases of study. The model for classifying the type of perturbations presents an accuracy of 98.31% with only 4 active detectors and 99.95% with 8 detectors; then it reaches the 100% of accuracy with 16 detectors active and is maintained for the other masks. However, the model for locating the perturbations has a lower accuracy for 4 active detectors, 90.35%, and the 99% accuracy is not reached until 16 detectors are active, reaching the 100% accuracy for 82 active detectors (Fig. 11).

These results also indicate that having all the detectors active does not improve the accuracy, with only 4 active detectors the accuracy reached is over 90% for both

Table 2 Percentage of accuracy for each 2D model and number of active detectors

Active detectors	Accuracy classification training (%)	Accuracy location training (%)
0	50.01	0.39
4	98.31	90.35
8	99.95	98.52
16	100	99.04
24	100	99.56
32	100	99.69
46	100	99.71
64	100	99.97
82	100	100
97	100	100
128	100	100
184	100	100
297	100	100

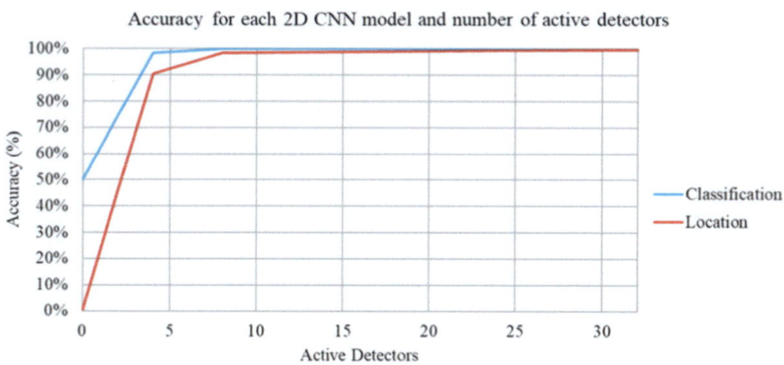

Fig. 11 Accuracy results for the 2D training models

trainings. What is more, with 8 active detectors the accuracy for the classification model surpasses 99% and for the location training the accuracy reaches percentages of 98.52%.

4.2 *Three-Dimensional IAEA Reactor*

The three-dimensional IAEA is represented in Fig. 12, and it is defined in [1]. The core consists of 177 fuel assemblies including nine fully rodded fuel assemblies and 9 partially rodded. The reactor is discretized in 19 axial planes. The size of each

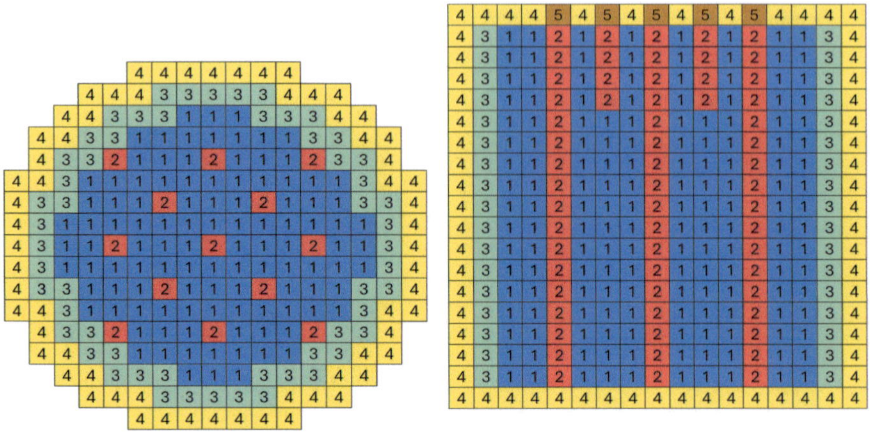

Fig. 12 Reactor geometry of the 3D IAEA benchmark

Table 3 Macroscopic cross sections of the IAEA 3D reactor

Material	D_1 (cm)	D_2 (cm)	Σ_{a1} (1/cm)	Σ_{a2} (1/cm)	Σ_{s12} (1/cm)	$\nu\Sigma_{f1}$ (1/cm)	$\nu\Sigma_{f2}$ (1/cm)
1–Fuel	1.50	0.40	0.010	0.085	0.020	0.00	0.135
2–Rodded fuel	1.50	0.40	0.010	0.130	0.020	0.00	0.135
3–Exterior fuel	1.50	0.40	0.010	0.080	0.020	0.00	0.135
4–Reflector	2.00	0.30	0.000	0.010	0.040	0.00	0.000
5–Rodded reflector	2.00	0.30	0.000	0.055	0.040	0.00	0.000

assembly is 20 × 20 × 20 cm. The cross-sections for its 5 homogenized materials with two-energy groups are given in Table 3. Albedo boundary conditions are used with an extrapolation distance of $2.13 \times D_G$.

Similarly to the 2D case of study, the three-dimensional models follow the architecture of the networks depicted in Figs. 8 and 9, considering the axial planes too. The quantity of the input files that feed the models increases, since a third type of perturbation, a change in the flow velocity, can occur. Then, it is important to develop an efficient model that can obtain high accuracies while investing a reasonable amount of computing time.

In Table 4, the results for the 3D training models are presented. These results for the accuracies show the same behaviour as the accuracies for the two-dimensional models. With more detectors active, the accuracies increase fast toward 100%, especially for the classification model. In the 3D reactor, the number of possible locations for the perturbation is increased enormously compared to the 2D geometry, which is reflected in the accuracies for locating the perturbation. The model for classifying the type of perturbations presents an accuracy of 99.11% with 8 active detectors and 99.57% with 16 detectors active, reaching the 100% of accuracy with the distribu-

Table 4 Percentage of accuracy for each 3D model and number of active detectors

Active detectors	Accuracy classification (%)	Accuracy location (%)
0	44.18	44.19
4	99.39	94.34
8	99.11	98.26
16	99.57	97.80
24	100	98.47
32	100	99.48
46	100	100
64	100	98.72
82	100	99.97
97	100	100
128	100	100
184	100	100
297	100	100

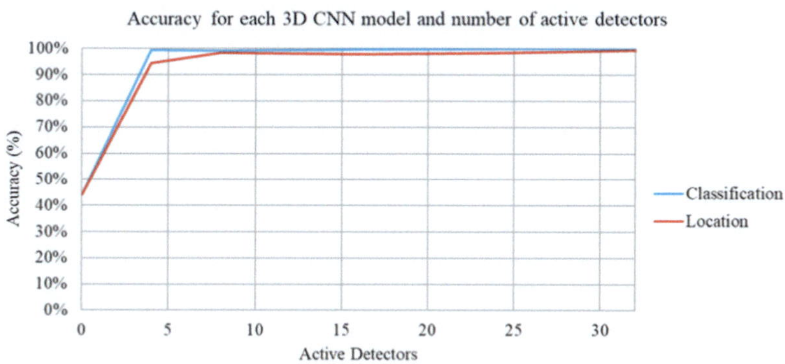

Fig. 13 Accuracy results for the 3D training models

tion of 24 active detectors, the 100% accuracy is maintained for the rest of masks of detectors. The three-dimensional training model for locating the perturbation present lower accuracies in comparison to the classifying model. With 8 active detectors the accuracy is 98.26% and the 99% of accuracy is not reached until 82 detectors are active, obtaining the 100% accuracy for 97 and more active detectors. However, the general tendency of both trainings is to obtain higher accuracies when the number of active detectors increases. The behaviour of the accuracy for these trainings is also shown in Fig. 13.

Then, choosing the minimum number of active detectors becomes crucial to optimize the calculation. It has been proved that using all the data from the simulations could lead to overfitting the training models as well as increase the calculation time without improving the results. Moreover, regarding the accuracy results of all the convolutional neural networks, the chosen architecture and the values of the hyper-parameters have proved to work well for our cases of study. Furthermore, the AWGN has been an especially useful tool to avoid overfitting while increasing the dataset without needing to generate new signals, together with enabling the possibility to introduce the errors during the core monitoring.

5 Conclusions

This work has presented the utility of convolutional neural networks to monitor the perturbations inside nuclear reactors, considering 2D and 3D geometries. The aim of the work is to obtain trained models that are capable of identify and locate the perturbations with high accuracy. To achieve the goal, a set of data for the BIBLIS 2D reactor and a set of signals for the 3D IAEA reactor are generated using FEMF-FUSION, considering the different types of perturbations and the geometry of the benchmarks.

However, a preprocessing of the data is needed to prepare the signals before they are used for the training. With this preprocessing it is assured that the training models would not overfit and that the convolutional networks would be optimal. On one hand, AWGN is added to the signals, creating new noisy signals to increase the dataset and avoid overfitting. On the other hand, 12 masks of active detectors are used to study the minimum number of detectors that need to be working to provide enough signals for the training models.

As is previously shown, the accuracy of the architectures proposed is directly dependent on the number of active detectors, improving the results when the number of active detectors increases. Nonetheless, in this work is considered that the optimal number of active detectors is 8, for the improvement of the results for more active detectors is not relevant in any of the trained models. Then, it is assumed that for the two-dimensional reactor, a 99.95% accuracy for identifying the type of perturbation and 98.52% accuracy are sufficient. Similarly, the three-dimensional reactor presents an accuracy of 99.11% for the identification model and an accuracy of 98.26% for locating the perturbations, which are also assumed to be enough.

In general terms, all the models present very high accuracies, especially due to the data preprocessing. With the original datasets, there is only one signal per type of perturbation, frequency of oscillation and position in the geometry, limiting the data for the training. After adding the AWGN the quantity of signals for training increases significantly, creating an extended variety of signals that are more balanced when they are split.

Therefore, it can be concluded that the results obtained are encouraging, proving that the signals preprocessing and the convolutional neural network models are suitable for the cases of study of this work. Then, this deep learning technique could be considered an efficient complement to the current in-core monitoring, contributing to maintaining the safe operation of the nuclear reactors.

In future works, validation with real data of nuclear plants is planned. Also, the simulation of more realistic perturbations like fuel assembly mechanical vibrations, core barrel vibrations, oscillations in the coolant flow (density and/or temperature) is scheduled.

Competing Interests The authors have no conflicts of interest to declare that are relevant to the content of this chapter.

Acknowledgements This research was supported by Universitat Politècnica de València and *Consejo de Seguridad Nuclear* under the project *INAIA (REF. SUB-34/2021)* and Cátedra CSN-UPV en Seguridad Nuclear y Protección Radiológica, developed in Instituto de Seguridad Industrial, Radiofísica y Medioambiental.

References

1. Argonne code center: benchmark problem book (1977). https://doi.org/10.2172/5037820. https://www.osti.gov/biblio/5037820
2. Calivá, F., De Ribeiro, F.S., Mylonakis, A., Demazière, C., Vinai, P., Leontidis, G., Kollias, S.: A deep learning approach to anomaly detection in nuclear reactors. In: 2018 International Joint Conference on Neural Networks (IJCNN), pp. 1–8 (2018). https://doi.org/10.1109/IJCNN.2018.8489130
3. Carreño, A., Vidal-Ferràndiz, A., Ginestar, D., Verdú, G.: Frequency-domain models in the spn approximation for neutron noise calculations. Prog. Nucl. Energy **148**, 104233 (2022). https://doi.org/10.1016/j.pnucene.2022.104233
4. Chollet, F., et al.: Keras (2015). https://keras.io
5. Ejigu, D.A., Liu, X.: Gradient descent-particle swarm optimization based deep neural network predictive control of pressurized water reactor power. Prog. Nucl. Energy **145**, 104108 (2022). https://doi.org/10.1016/j.pnucene.2021.104108
6. Fontenla, Y., Vidal-Ferràndiz, A., Carreño, A., Ginestar, D., Verdú, G.: Femffusion and its verification using the c5g7 benchmark. Ann. Nucl. Energy **196**, 110239 (2024). https://doi.org/10.1016/j.anucene.2023.110239. http://dx.doi.org/10.1016/j.anucene.2023.110239
7. Huang, L., Pan, W., Zhang, Y., Qian, L., Gao, N., Wu, Y.: Data augmentation for deep learning-based radio modulation classification. IEEE Access **8**, 1498–1506 (2020). https://doi.org/10.1109/access.2019.2960775
8. Hébert, A.: Application of the hermite method for finite element reactor calculations. Nucl. Sci. Eng. **91**(1), 34–58 (1985). https://doi.org/10.13182/nse85-a17127
9. Kamkar, A., Abbasi, M.: A comparative study of machine learning approaches for identification of perturbed fuel assemblies in wwer-type nuclear reactors. Ann. Nucl. Energy **211**, 110992 (2025). https://doi.org/10.1016/j.anucene.2024.110992. http://dx.doi.org/10.1016/j.anucene.2024.110992
10. Kollias, S., Yu, M., Wingate, J., Durrant, A., Leontidis, G., Alexandridis, G., Stafylopatis, A., Mylonakis, A., Vinai, P., Demaziere, C.: Machine learning for analysis of real nuclear plant data in the frequency domain. Ann. Nucl. Energy **177**, 109293 (2022). https://doi.org/10.1016/j.anucene.2022.109293
11. Müller, E., Weiss, Z.: Benchmarking with the multigroup diffusion high-order response matrix method. Ann. Nucl. Energy **18**(9), 535–544 (1991). https://doi.org/10.1016/0306-4549(91)90098-i

12. Müller, E.Z., Weiss, Z.J.: Benchmarking with the multigroup diffusion high-order response matrix method. Ann. Nucl. Energy **18**(9), 535–544 (1991). https://doi.org/10.1016/0306-4549(91)90098-I
13. Mylonakis, A., Vinai, P., Demazière, C.: Core sim+: a flexible diffusion-based solver for neutron noise simulations. Ann. Nucl. Energy **155**, 108149 (2021). https://doi.org/10.1016/j.anucene.2021.108149. http://dx.doi.org/10.1016/j.anucene.2021.108149
14. Papaoikonomou, A., Wingate, J., Verma, V., Durrant, A., Ioannou, G., Papagiannis, T., Yu, M., Alexandridis, G., Dokhane, A., Leontidis, G., Kollias, S., Stafylopatis, A.: Deep learning techniques for in-core perturbation identification and localization of time-series nuclear plant measurements. Ann. Nucl. Energy **178**, 109373 (2022). https://doi.org/10.1016/j.anucene.2022.109373
15. Pedregosa, F., Varoquaux, G., Gramfort, A., Michel, V., Thirion, B., Grisel, O., Blondel, M., Prettenhofer, P., Weiss, R., Dubourg, V., Vanderplas, J., Passos, A., Cournapeau, D., Brucher, M., Perrot, M., Duchesnay, E.: Scikit-learn: machine learning in Python. J. Mach. Learn. Res. **12**, 2825–2830 (2011). https://doi.org/10.5555/1953048.2078195
16. Pázsit, I., Demazière, C.: Noise Techniques in Nuclear Systems, pp. 1629–1737. Springer US (2010). https://doi.org/10.1007/978-0-387-98149-9_14
17. Simonyan, K., Zisserman, A.: Very deep convolutional networks for large-scale image recognition (2014). arXiv:1409.1556
18. Tasakos, T., Ioannou, G., Verma, V., Alexandridis, G., Dokhane, A., Stafylopatis, A.: Deep learning-based anomaly detection in nuclear reactor cores. In: Proceedings of the International Conference on Mathematics & Computational Methods Applied to Nuclear Science & Engineering (M&C 2021), Online, pp. 3–7 (2021)
19. Taylor, L., Nitschke, G.: Improving deep learning with generic data augmentation. In: 2018 IEEE Symposium Series on Computational Intelligence (SSCI), pp. 1542–1547. IEEE (2018)
20. Vidal-Ferràndiz, A., Carreño, A., Ginestar, D., Verdú, G.: FEMFFUSION: A finite element method code for the neutron diffusion equation (2022). https://femffusion.webs.upv.es/
21. Vidal-Ferràndiz, A., Carreño, A., Ginestar, D., Verdú, G.: Edge-wise perturbations to model vibrating fuel assemblies in the frequency-domain using femffusion: Development and verification. Ann. Nucl. Energy **175**, 109246 (2022)
22. Vidal-Ferràndiz, A., Carreño, A., Ginestar, D., Demazière, C., Verdú, G.: A time and frequency domain analysis of the effect of vibrating fuel assemblies on the neutron noise. Ann. Nucl. Energy **137**, 107076 (2020). https://doi.org/10.1016/j.anucene.2019.107076

A New Proposal for Team Ranking in First Division of Spanish Football League

Francisco Pedroche

Abstract We introduce a novel methodology for ranking teams in the first division of the Spanish football league. Given the outcomes of the fixtures in a match-week (round or *jornada* in Spanish), our approach ranks teams based on three criteria: two dependent on the results of the fixtures (goal difference and scored or conceded goals) and a third feature employed as a tie-breaking rule, namely each team's position in the preceding round's ranking. Upon ranking the teams, our method assigns 25, 18, 15, 12, 10, 8, 6, 4, 2, and 1 point, respectively, to the top ten teams. This yields a ranking associated with each round and, by accumulating these points across consecutive rounds, produces a final classification for the entire season. We apply our methodology to the 2022–2023 and 2023–2024 championships and utilize a measure of Competitive Balance to compare the resulting series of rankings with the official ones. For the studied seasons, our proposed method exhibits a higher Competitive Balance than the official system. Furthermore, our approach selects the same teams for the Champions League as the official method. The method is likely to increase fan interest by offering teams more opportunities to attain top positions.

Keywords Kendall's tau · Competitive Balance · Sport rankings · Football statistics

1 Introduction

Sport rankings serve as a systematic framework, enabling an objective assessment of team performances whilst delineating their comparative standing within the competition. Such evaluations offer a quantifiable metric of success that significantly influences both the financial health of clubs and their capacity to captivate audiences. With specific reference to football (soccer) team rankings, the primary data for such rankings is derived from league standings. For instance, in the con-

F. Pedroche (✉)
Institut Universitari de Matemàtica Multidisciplinària, Universitat Politècnica de València, Valencia, Spain
e-mail: pedroche@mat.upv.es

J. R. Torregrosa (ed.), *Mathematical Modelling in Engineering Human Behaviour*, Springer Proceedings in Mathematics & Statistics 513,
https://doi.org/10.1007/978-3-032-00567-0_3

Table 1 Top 6 standings for the official 2023–2024 LFP along the last seven rounds

R. 32	R. 33	R. 34	R. 35	R. 36	R. 37	R. 38
R. Madrid	R. Madrid	R. Madrid	R. Madrid	R. Madrid	R. Madrid	R. Madrid
Barcelona	Barcelona	Girona	Barcelona	Barcelona	Barcelona	Barcelona
Girona	Girona	Barcelona	Girona	Girona	Girona	Girona
Atlético	Atlético	Atlético	Atlético	Atlético	Atlético	Atlético
Athletic	Athletic	Athletic	Athletic	Athletic	Athletic	Athletic
R. Sociedad	R. Sociedad	R. Sociedad	Betis	R. Sociedad	R. Sociedad	R. Sociedad

text of Spain's premier division -officially termed *Campeonato Nacional de Liga Profesional Primera División*, hereafter referred to as LFP first division or simply LFP- pertinent information can be accessed via reference [21]. The commercial name for the championship is LALIGA. LALIGA is a private sports association composed of the 20 clubs and SADs (public limited sports companies) in LALIGA EA SPORTS (first division) and the 22 in LALIGA HYPERMOTION (second division) [23].

The LFP first division (and also the second division) is organised as a double round robin tournament, i.e., a tournament in which each team plays each other team twice, once at home and once away. The games of the tournament are grouped in rounds (matchweeks or *jornadas*, in Spanish), such that the team plays exactly once per round. Normally each round corresponds to a week-end but sometimes one week can hold two rounds, as happened recently in rounds 3 and 4 of session 24–25 of the first division. Since the season 1997–1998, 20 teams take part in this competition.

The classification within the LFP is determined considering a multitude of factors. Chief among these are points accrued (three for a win, one for a draw, and none for a loss), goal difference, total goals scored, and goals conceded (see [22]). A notable characteristic of the LFP ranking's evolution over time is that certain teams maintain a leading position over extended periods. For example, Table 1 depicts the performance of the top six teams across the final seven weeks of the 2023–2024 LFP championship. This table reveals Real Madrid's steadfast occupancy of first place throughout this period while FC Barcelona and Girona FC experienced two instances of positional crossover (ranging from week 33 to week 35).

Note that the LFP championship can be considered as a series of rankings, and this is the main point of view that we hold in this communication. In more detail, we use the Kendall's τ coefficient of disagreement (see [13–15]) to measure the evolution from ranking to ranking and the *evolutive Kendall's correlation coefficient* τ_e, introduced in [4], to assign a measure of *Competitive Balance* to a series of rankings. This allows us to compare our series of rankings with the corresponding rankings given by the official LFP classifications.

In a recent work (see [17]) we compared the competitiveness of the Formula one (F1) championship and the LFP championship along ten years by using the *Corrected Evolutive Kendall's tau* coefficient that was introduced in [18]. As one of the conclusions we remarked that the F1 system was more captivating than the

LFP system because there were more movements in the Grand Prix (GP) standings (virtually any pilot can win a GP). As a consequence we recommended to follow a similar system for the LFP. That is, a ranking system that will allow any team to be on the first place, any week. Guided by this idea, in this communication we present a method to rank the LFP in such a way that any round we produce a new ranking. In addition, by accumulating the points assigned in each round, we also produce a parallel general classification to be able to give a final ranking of the championship. To support the validity of the proposed method we also compare the final ranking of our method with the official final ranking given by the LFP classification system.

The structure of the paper is the following. In Sect. 1.1 we give some references and recall some definitions and results to measure competitive balance in sports, focusing mainly on football, and we give some examples to illustrate the introduced concepts. In Sect. 2 we describe our proposed method to rank the teams each round of the LFP championship. In Sect. 2.2 we introduce an scheme to share points each round, inspired by the F1 championship, and we show how to obtain a general classification based on accumulating these points along the season. In Sect. 3 we apply the new method to the seasons 2022–2023 and 2023–2024 of the LFP first division, and we compare the orders and the movements given by our series of rankings with the official LFP ones. Finally, in Sect. 4 we outline some conclusions of the study.

1.1 Kendall's Tau as a Measure of Competitive Balance

Competitiveness is one of the major features required to a sport contest. The usual way to refer to competitiveness in Sports Economics literature is by using the term competitive balance (hereinafter CB). In short, it is said that a sport contest or a league championship has very low CB when there is a team or a small group of teams that always win. In such a case, it is very easy to guess who wins and this fact diminishes fan interest. There is some empirical evidence that even the fans of winnings teams lose interest when the outcome of a match is very predictable (see [8, 12]). Therefore, it is usually assumed that the higher the uncertainty in the result (or similarly, the higher the equality among the teams), the higher fan interest, and the higher CB; this is known as the "Uncertainty of Outcome Hypothesis" (UOH) of Rottenberg (see [20]). However, there is a great debate in the literature about what is the precise definition of CB; see, for example, the discussion in [10].

Sports leagues are usually managed by private or public institutions with the objective of increasing their incomes. In professional sports, revenues are mainly due to gate attendances, merchandising, media, sponsorship, etc. Therefore, sport leagues seek to offer matches with high equality among the teams, with high CB, to attract fans; see, e.g., [1, 6, 8, 16] for details on how to measure CB on football.

An alternative approach to quantifying competitive balance (CB) involves calculating an index that captures the mobility of teams across seasons [12], or rounds [4]. This metric leverages Kendall's rank correlation coefficient, providing a nuanced assessment of team performance variability over time. As demonstrated in [14],

Kendall's τ coefficient–also referred to as the measure of disarray–associated with two rankings comprising an equal number of elements, n, can be expressed in the form

$$\tau = 1 - \frac{2s}{\frac{1}{2}n(n-1)} \tag{1}$$

where s represents the minimal number of pairwise exchanges necessary to transform one ranking into another. This coefficient serves as a quantitative indicator of the strength of rank correlation, thereby facilitating the assessment of the underlying relationship between the two rankings. Note that $\tau \in [-1, 1]$. The coefficient can also be written as

$$\tau = \frac{P - Q}{\frac{1}{2}n(n-1)} \tag{2}$$

where P denotes the number of element pairs that preserve their relative ordering when transitioning from the initial ranking to the subsequent one (i.e., the first element remains above or below the second in both rankings), and Q represents the number of element pairs that reverse their order (i.e., the first element is above the second in one ranking, while being below it in the other, or vice versa). Notably, Q and s are equivalent quantities. Moreover, this metric can be interpreted as the number of crossings or inversions (see, e.g. [3, 7]) that occur when transitioning from the initial ranking to the subsequent one.

Example 1
Let the ordered set $c_t = \{1, 2, \dots, n\}$ be the ranking (*standings or classification*) of n teams at round t. Note that each number is a tag to identify each team. Let c_{t+1} be the ranking at the next round. Obviously, c_{t+1} is a permutation of c_t. Let us denote this permutation by π, and write $\pi c_t = c_{t+1} \equiv \{i_1, i_2, \dots i_n\}$. If the team tagged as j goes to position i_k we write $\pi(j) = i_k$. We say that the teams i and j have crossed their positions from c_t to c_{t+1} if $i < j$ and $\pi(i) > \pi(j)$. For example, let $c_t = \{1, 2, 3, 4\}$ and $c_{t+1} = \{3, 2, 4, 1\}$. Therefore $\pi(1) = 4, \pi(2) = 2, \pi(3) = 1, \pi(4) = 3$. The teams 1 and 4 have crossed their positions since $1 < 4$ and $\pi(1) > \pi(4)$. The teams 2 and 4 have not crossed their positions. It is easy to see that the total number of crossings is $s = 4$ and therefore, from Eq. 1 we have that $\tau = -1/3$.

In [4] two teams that have crossed their positions are said to *compete* with each other. It is also interesting to recall, from [9], that a *distance metric* $d(\mathbf{a}, \mathbf{b})$ can be related to a correlation coefficient $\tau(\mathbf{a}, \mathbf{b})$ by

$$\tau(\mathbf{a}, \mathbf{b}) = 1 - \frac{2d(\mathbf{a}, \mathbf{b})}{d_{max}(\mathbf{a}, \mathbf{b})} \tag{3}$$

where $d_{max}(\mathbf{a}, \mathbf{b})$ is the maximum possible distance between two rankings[1]. Note that Eq. (1) is of this form, since $n(n-1)/2$ is the maximum number of crossings between two given rankings.

For the case of n consecutive rankings, in [4] an evolutive Kendall's correlation coefficient τ_e for a set of r rankings $R = \{c_1, c_2, \ldots, c_r\}$ is introduced in the form

$$\tau_e = 1 - 4 \frac{\omega(R)}{n(n-1)(r-1)} \tag{4}$$

where $\omega(R)$ is the number of accumulative crossings between consecutive rankings and n is the number of elements of each ranking. The authors introduce this concept in the framework of a *Competitivity Graph* that captures the crossings in the standings throughout a season, and some CB measures are proposed. In particular, they apply this measure for analyzing the four major European football leagues (England, Germany, Italy and Spain), obtaining that the Bundesliga was the most competitive in season 2013–2014. For a comparison of some classical measures of CB and τ_e, see [19].

Another quantity (borrowed from graph theory) is also introduced in [4]: the *Normalized Mean Strength* NS; that is, the normalized sum of the weights of the edges of a weighted graph. When considering a series of r consecutive rankings of n elements and its corresponding *Competitivity Graph*, we have the following relation

$$NS = \frac{1 - \tau_e}{2} \tag{5}$$

that gives the equivalence between the Kendall's τ_e and the *Normalized Mean Strength* for r rankings. Note that $\tau_e \in [-1, 1]$ and $NS \in [0, 1]$. We contend that the measure NS offers a more intuitive understanding compared to τ_e, as it enables the interpretation of movements or *activity* within a sequence of rankings as a percentage, thereby facilitating a more accessible and comprehensible analysis.

Example 2
Let us consider three rankings of four elements: $c_1 = \{1, 2, 3, 4\}$, $c_2 = \{3, 2, 4, 1\}$, and $c_3 = \{2, 3, 1, 4\}$. It is easy to see that the number of crossings from c_1 to c_2 is 4, and the number of crossings from c_2 to c_3 is 2, therefore the total number of crossings is $\omega(c_1, c_2, c_3) = 6$ and therefore, from (4), $\tau_e = 0$, and from (5), $NS = 0.5$.

In summary, we will employ the quantity NS as a metric for quantifying competitive balance (CB) across r rankings. By leveraging this measure, we will conduct a comparative analysis of the competitiveness of a championship under two

[1] A distance metric between two rankings **a** and **b** is a nonnegative real function f that is *symmetric* ($f(\mathbf{a}, \mathbf{b}) = f(\mathbf{b}, \mathbf{a})$, for any pair of rankings), *regular* ($f(\mathbf{a}, \mathbf{b}) = 0 \leftrightarrow \mathbf{a} = \mathbf{b}$) and satisfying the *triangle inequality* ($f(\mathbf{a}, \mathbf{c}) \leq f(\mathbf{a}, \mathbf{b}) + f(\mathbf{b}, \mathbf{c})$, for any rankings **a**, **b**, and **c**).

distinct schemes for defining rankings: specifically, when utilizing consecutive rankings given by the official classification provided by the LFP, and when applying our novel method to produce consecutive rankings.

2 Methodology

In this section, we present a novel methodology for ranking teams at each round of a championship. We give the details about each step of the computations and give an algorithmic code of the computation for an illustrative round. In addition, we show the applicability of our method through some examples. Furthermore, we propose an approach to derive a comprehensive final classification for the entire championship, providing a holistic perspective on team performance, that is not so different (as we will show in Sect. 3) of the official final classification.

2.1 Proposed Method for Obtaining a Ranking for Each Round

Consider the LFP championship, comprising 38 rounds. For the purpose of exposition, assume that the outcomes of all fixtures for each round are known (i.e., the championship has concluded). Our proposed methodology operates as follows: at each round, we generate a ranking of the 20 participating teams according to the following **Ranking Team Rules**:

1. Initial Classification: The 20 teams are initially categorized into three distinct groups: winners, tied teams, and losers, according to their respective match outcomes.
2. Intra-Group Ranking: Within each group, teams are ordered based on their goal difference (GD), with higher positions allocated to teams exhibiting larger GD values. In the event of a tie in GD between two or more teams, the following secondary criteria are applied:
   - The team scoring the most goals is assigned a higher position.
   - If the tie persists, the team with a superior ranking in the previous round is preferred.
   - Exception for the First Round: In the inaugural week of the championship, the preceding rank refers to the final classification of the previous LFP championship. This consideration ignores teams relegated to the second division and incorporates teams promoted to the first division.
3. Final Ranking Composition: The three rankings obtained through the aforementioned steps are subsequently concatenated, with precedence given to winner teams, followed by tied teams, and lastly, loser teams.

Table 2 Ranking r_0 for season 2023–2024

Position	Club
1	Barcelona
2	Real Madrid
3	Atlético
4	R. Sociedad
5	Villarreal
6	Betis
7	Osasuna
8	Athletic
9	Mallorca
10	Girona
11	Rayo
12	Sevilla
13	Celta
14	Cádiz
15	Getafe
16	Valencia
17	Almería
18	Granada
19	Las Palmas
20	Alavés

The next example illustrates the proposed method.

Example 3
Let us show how to apply this method to compute the first ranking corresponding to the LFP season 2023–2024. We first need to elaborate the rank r_0 corresponding to the previous championship, replacing the descended teams by the promoted ones. The corresponding ranking is shown in Table 2, while the results of the fixtures corresponding to the first round are shown in Table 3. From Table 3 we create the following groups:

Winners: Rayo, Valencia, R. Madrid, Osasuna, Betis, Cádiz, Atlético.
Tied: R. Sociedad, Girona, Las Palmas, Mallorca, Getafe, Barcelona.
Losers: Almería, Sevilla, Athletic, Celta, Villarreal, Alavés, Granada

Now we rank the teams in each group attending the goal difference in the match. We break the ties by given preference to the team with a greater number of scored goals in its match. We obtain:

Table 3 Results from round 1, season 2023–2024, LFP first division

Home team	Goals	Away team	Goals
Almería	0	Rayo	2
Sevilla	1	Valencia	2
R. Sociedad	1	Girona	1
Las Palmas	1	Mallorca	1
Athletic	0	R. Madrid	2
Celta	0	Osasuna	2
Villarreal	1	Betis	2
Getafe	0	Barcelona	0
Cádiz	1	Alavés	0
Atlético	3	Granada	1

Winners: Atlético, Rayo, R. Madrid, Osasuna, Betis, Valencia, Cádiz.
Tied: R. Sociedad, Girona, Las Palmas, Mallorca, Getafe, Barcelona.
Losers: Sevilla, Villarreal, Alavés, Granada, Almería, Athletic, Celta.

Finally we break the ties, in each group, by looking at the ranking obtained in r_0. We get:

Winners: Atlético, R. Madrid, Osasuna, Rayo, Betis, Valencia, Cádiz,
Tied: R. Sociedad, Mallorca, Girona, Las Palmas, Barcelona, Getafe.
Losers: Villarreal, Sevilla, Alavés, Granada, Athletic, Celta, Almería.

Lastly, we only have to stack these three rankings to obtain the final ranking produced by the method corresponding to the round 1. We show this ranking in Table 4.

In the practice, the method can be simplified by using sorting strategies of programming languages. For example, by defining a vector `DG` that contains the goal difference for each team (goals scored minus goals conceded) and a vector `G` with the goals scored by each team we can easily write the procedure shown in Example 3 in some Python lines, as shown in Example 4. In this way, the value of `sorted_df` is precisely the ranking r_1 shown in Table 4.

Example 4. Python code for computing r_1 for LFP season 23–24

```
import pandas as pd
import numpy as np
data =
{'Name':['Barcelona', 'Real Madrid', 'Atletico',
        'R. Sociedad','Villarreal','Betis',
```

Table 4 Ranking r_1 for season 2023–2024 with the new method

Position	Club
1	Atlético
2	Real Madrid
3	Osasuna
4	Rayo
5	Betis
6	Valencia
7	Cádiz
8	R. Sociedad
9	Mallorca
10	Girona
11	Las Palmas
12	Barcelona
13	Getafe
14	Villarreal
15	Sevilla
16	Alavés
17	Granada
18	Athletic
19	Celta
20	Almería

```
         'Osasuna', 'Athletic', 'Mallorca',
         'Girona', 'Rayo', 'Sevilla','Celta',
         'Cadiz', 'Getafe','Valencia', 'Almeria',
         'Granada', 'Las Palmas', 'Alaves'],
   'Tag': np.arange(20),
   'DG':[0,2,2,0,-1,1,2,-2,0,0,2,-1,-2,1,0,1,-2,-2,0,-1],
   'G':[0,2,3,1,1,2,2,0,1,1,2,1,0,1,0,2,0,1,1,0]}
   df = pd.DataFrame(data)
   sorted_df = df.sort_values(['DG', 'G','Tag'],
                       ascending=[False, False, True])
```

It is noteworthy that the algorithmic implementation of our proposed method for each round of the entire championship involves an iterative application of the code exemplified in Example 4. This process entails modifying the input ranking, denoted by the column 'Name', which is copied from the preceding ranking. Furthermore, the columns `'DG'` and `'G'` are recomputed according to the outcomes of the matches

Table 5 Points scoring sharing

1st	2nd	3rd	4th	5th	6th	7th	8th	9th	10th
25	18	15	12	10	8	6	4	2	1

in the current round. To illustrate this procedure, consider the computation of the ranking corresponding to round 2. This involves executing a code similar to that in Example 4, but with the following adaptations:

- Replace the list 'Name' with the ranking r_1 presented in Table 4.
- Update the vectors 'DG' and 'G' with the respective values according to the results of the fixtures that took place in round 2.

2.2 *Proposed Method for Obtaining a Final Ranking*

As highlighted in the introduction, our novel method draws inspiration from the Formula 1 (F1) championship. Since our approach generates a ranking for each round, we can assign a quantity of points to the top ten ranked teams, inspired by the F1 championship's point allocation system, as depicted in Table 5. By accumulating the points obtained by each team in each round, we can establish a comprehensive general classification for the entire championship. This procedure is exemplified in Example 5, which demonstrates the assignment of points and the subsequent accumulation to determine the overall championship standings.

Example 5. General classification corresponding to round 1, LFP 23–24
The first ten teams have the following scores: Atlético (25), Real Madrid (18), Osasuna (15), Rayo (12), Betis (10), Valencia (8), Cádiz (6), R. Sociedad (4), Mallorca (2), and Girona (1). The rest of the teams do not get any points in round 1, and their rankings are the ones shown in Table 4.

3 Results

In this section, we present the application of our novel methodology to two seasons of the first division LFP championship: 2022–2023 and 2023–2024. For each season, we juxtapose the evolution of the rankings generated by our new method with those obtained from the official LFP tables, thereby facilitating a comparative analysis of the two approaches.

Table 6 Ranking r_0 for season 2022–2023

Position	Club
1	Real Madrid
2	Barcelona
3	Atlético
4	Sevilla
5	Betis
6	R. Sociedad
7	Villarreal
8	Athletic
9	Valencia
10	Osasuna
11	Celta
12	Rayo
13	Elche
14	Espanyol
15	Getafe
16	Mallorca
17	Cádiz
18	Almería
19	Real Valladolid
20	Girona

3.1 LFP Season 2022–2023

As we have pointed out before, the starting point of the proposed method is the ranking r_0 corresponding to the previous season, replacing the descended teams by the promoted ones. The ranking r_0 to be used as starting point for the season 2022–2023 is shown in Table 6. This table will be used as a tie-breaking rule if some ties are encountered in the step 2 of the proposed method. By using the results of the fixtures of the 38 rounds of the season (that can be accessed in [21]) we produced the 38 rankings given by our method. As an example, we show in Fig. 7 the last six rankings produced by our method. To highlight the differences between our method and the official method we show in Table 8 the last six rankings produced by the official LFP method. Initially, a cursory examination reveals that the two sequences of rankings exhibit striking disparities. Specifically, whereas one sequence displays considerable variability in its rankings, the other sequence exhibits a notable phenomenon wherein the top six positions are predominantly occupied by a select group of six teams. A quantified analysis of these findings is presented in the subsequent sections.

Table 7 Standings computed with the proposed method for the last six weeks of 2022–2023 LFP

W. 33	W. 34	W. 35	W. 36	W. 37	W. 38
Atlético	Villarreal	Atlético	R. Valladolid	Barcelona	Mallorca
R. Sociedad	Sevilla	Almería	Villarreal	Real Madrid	Osasuna
Sevilla	Barcelona	Cádiz	Osasuna	Getafe	R. Sociedad
Girona	Betis	Villarreal	Real Madrid	Atlético	Celta
Almería	Osasuna	R. Sociedad	R. Sociedad	Rayo	Espanyol
Rayo	Valencia	Espanyol	Getafe	Betis	Almería
Barcelona	Elche	Athletic	Mallorca	Elche	Atlético
Getafe	Mallorca	Valencia	Atlético	Cádiz	Villarreal
Betis	Real Madrid	Elche	Espanyol	Espanyol	Real Madrid
Villarreal	R. Sociedad	Getafe	Elche	Valencia	Betis
Valencia	Girona	Sevilla	Sevilla	R. Valladolid	Elche
Espanyol	Celta	Betis	Girona	Almería	Cádiz
Elche	Atlético	Barcelona	Celta	Villarreal	Valencia
Mallorca	Getafe	Girona	Rayo	Osasuna	Athletic
R. Valladolid	Cádiz	Celta	Almería	R. Sociedad	Getafe
Athletic	Espanyol	Rayo	Valencia	Sevilla	R. Valladolid
Celta	Almería	Real Madrid	Betis	Girona	Barcelona
Osasuna	Rayo	R. Valladolid	Barcelona	Celta	Sevilla
Real Madrid	R. Valladolid	Osasuna	Cádiz	Athletic	Girona
Cádiz	Athletic	Mallorca	Athletic	Mallorca	Rayo

3.1.1 Comparison of Season Competitiveness as Measured by NS

By using the measure NS we can assign a competitiveness index to a season. In Table 9 we show the values of NS corresponding to the series of rankings produced by the proposed method and by using the official rankings for season 2022–2023. In [17] it is shown that the values of NS for the LFP seasons 2012–2013 to 2022–2023, computed by using the official rankings, range from 0.044 to 0.076, with a mean value around 0.06. As a conclusion from Table 9 we observe that the value of NS given by the novel method is much bigger than the one corresponding to the series given by the official LFP system. Therefore, our method gives more CB to the championship than the official method.

3.1.2 Comparative Analysis of Teams Obtaining Victories in Individual Rounds

Figure 1 presents a juxtaposition of the teams that attain the top ranking in the official LFP standings for the 2022–2023 season, alongside those that emerge as leaders when

Table 8 Official standings for the last six weeks of 2022–2023 LFP

W. 33	W. 34	W. 35	W. 36	W. 37	W. 38
Barcelona	Barcelona	Barcelona	Barcelona	Barcelona	Barcelona
Atlético	Real Madrid	Atlético	Real Madrid	Real Madrid	Real Madrid
Real Madrid	Atlético	Real Madrid	Atlético	Atlético	Atlético
R. Sociedad	R. Sociedad	R. Sociedad	R. Sociedad	R. Sociedad	R. Sociedad
Villarreal	Villarreal	Villarreal	Villarreal	Villarreal	Villarreal
Betis	Betis	Betis	Betis	Betis	Betis
Girona	Girona	Athletic	Osasuna	Osasuna	Osasuna
Athletic	Athletic	Girona	Athletic	Athletic	Athletic
Rayo	Osasuna	Sevilla	Girona	Girona	Mallorca
Osasuna	Sevilla	Osasuna	Sevilla	Rayo	Girona
Sevilla	Rayo	Rayo	Mallorca	Sevilla	Rayo
Mallorca	Mallorca	Mallorca	Rayo	Mallorca	Sevilla
Celta	Celta	Valencia	Valencia	Cádiz	Celta
Almería	Valencia	Almería	Celta	Getafe	Cádiz
Cádiz	Almería	Celta	Almería	Valencia	Getafe
R. Valladolid	Cádiz	Cádiz	Getafe	Almería	Valencia
Valencia	R. Valladolid	Getafe	Cádiz	Celta	Almería
Getafe	Getafe	R. Valladolid	R. Valladolid	R. Valladolid	R. Valladolid
Espanyol	Espanyol	Espanyol	Espanyol	Espanyol	Espanyol
Elche	Elche	Elche	Elche	Elche	Elche

Table 9 Values of NS

Season	LFP	New method
2022–2023	0.059	0.462

employing the proposed methodology for the same seasonal period. We observe that only four teams accessed to the first place with the official method, while 15 teams accessed the first position with the novel proposal.

3.1.3 Comparison of Runner-Up Teams in Individual Rounds

In Fig. 2 we show the teams that occupy the second position in the official LFP rankings for the season 2022–2023, and the teams that occupy the same position by using the proposed method for the same seasonal period. Note that only four teams achieved the second place with the official method, while 15 teams accessed the same position with the novel proposal.

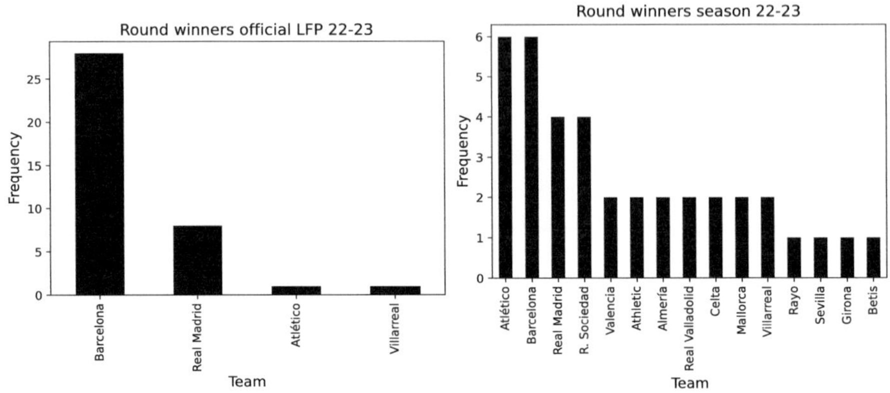

Fig. 1 Teams that occupy the first position at any round in the official rankings (left panel) and by the proposed method, in season LFP 2022–2023

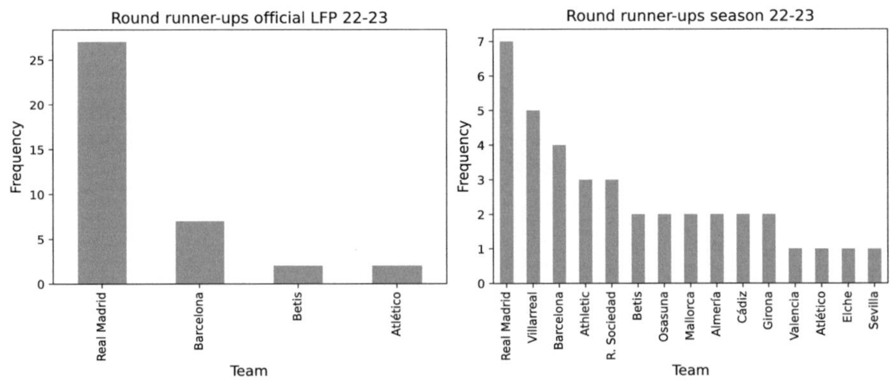

Fig. 2 Teams that occupy the second position at any round in the official (left panel) and by the proposed method, in season LFP 2022–2023

3.1.4 Comparison of Teams Occupying the Third Position in Individual Rounds

In Fig. 3 we show the teams that occupy the third position in the official LFP rankings for the season 2022–2023 (left panel), and the teams that occupy the same position when using the rankings computed by the new method. Note that only seven teams achieved the third place with the official method, while 15 teams accessed the same position with the new proposal.

In the next section we will direct our attention to the subset of teams that have simultaneously occupied the first, second, and third positions in any of the standings within the same championship. Specifically, we seek to identify the intersection of the sets of teams that emerged in the preceding subsections, namely those that secured top rankings in each of the three categories.

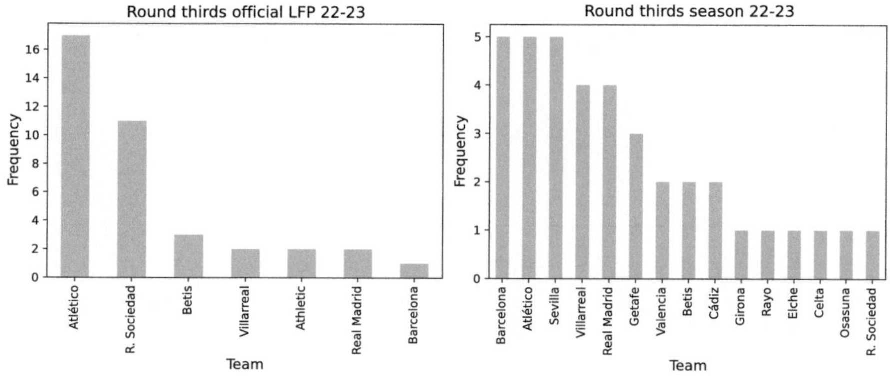

Fig. 3 Teams that occupy the third position at any round in the official (left panel) and by the proposed method, in season LFP 2022–2023

3.1.5 Comparative Examination of Teams Occupying the Top 3 Positions

Last column in Table 8 reveals that the Top 3 teams according to the official LFP classification for the 2022–2023 season comprise FC Barcelona, Real Madrid, and Atlético de Madrid. The frequency with which these teams occupy the first, second, and third positions in the official standings is illustrated in Fig. 4 (upper panel). As observed, FC Barcelona secured the top ranking during 28 weeks, whereas Real Madrid only achieved this feat on 8 occasions. In contrast, Fig. 4 (lower panel) displays the teams that occupied the Top 3 positions when employing our proposed methodology. This figure highlights a notable distinction, wherein our approach yields a more diverse set of winning teams.

Another form to compare the movement of the teams along the rankings is by focusing only on the Top 3 teams. We do this in Fig. 5 where we compare the movements along the classifications of FC Barcelona, Real Madrid and Atlético de Madrid when using the official classifications (upper panel) and when using the rankings computed by the proposed method (lower panel).

3.1.6 General Classification for LFP 2022–2023

As indicated in Sect. 2.2, by accumulating the points obtained by each team in each round, we can establish a comprehensive general classification for the entire championship. In each round the points are given following the scheme that was shown in Table 5. For the season LFP 2022–2023 the general classification derived from the novel method is shown in Table 10 jointly with the official classification for that season. Some features can be remarked from this table. First, the teams that qualify for the competition UEFA Champions League are the same in both classifications,

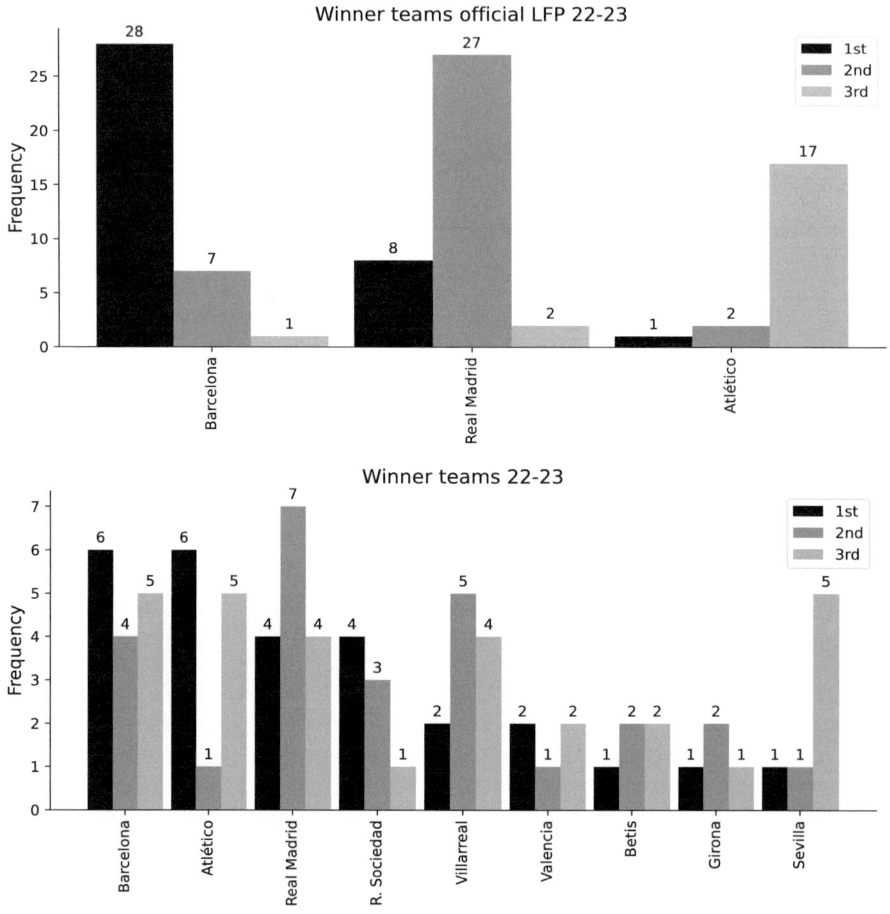

Fig. 4 Times that the Top 3 teams achieved the first, second and third place in the official rankings LFP 2022–2023 (up) and when using the rankings computed by the proposed method (down)

because this selected group is made of the four best classified teams. Second, the teams that qualify for the UEFA Europa League are also the same in both systems (fifth and sixth ranked teams). However the systems do not coincide for the team qualified for the UEFA Conference League championship (seventh ranked team). A third feature that can be remarked is that the systems differ slightly in which teams must be relegated (the three last teams). We consider that these similarities with the official system serve as a proof of the consistency of our novel method (that is not mainly devoted to give a final classification but to give a novel classification for each round).

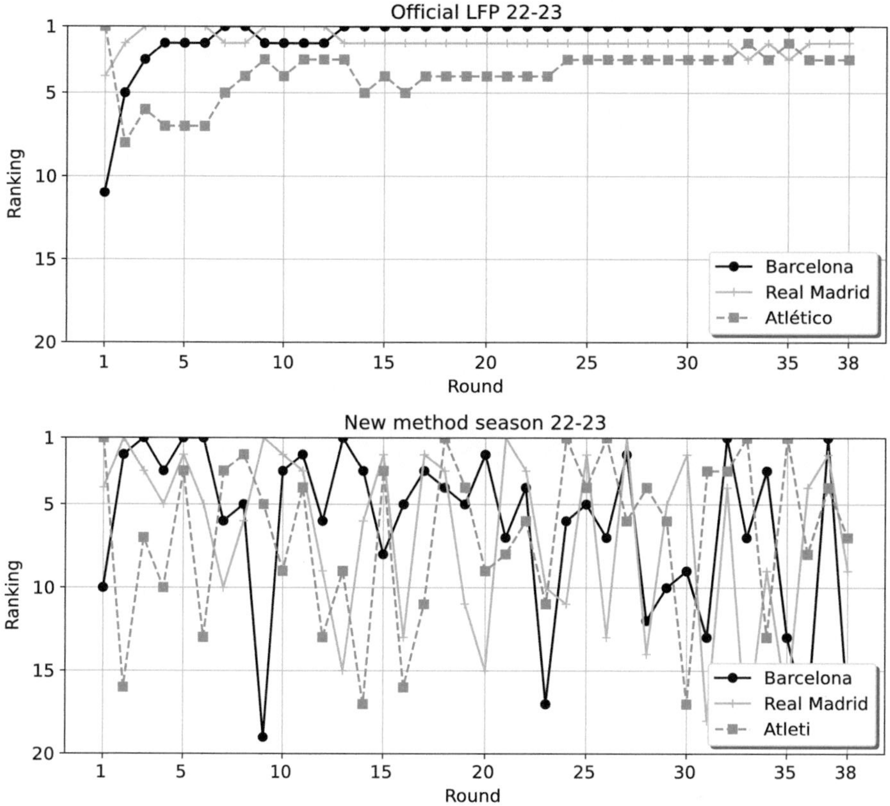

Fig. 5 Movements along the classifications of FC Barcelona, Real Madrid and Atlético de Madrid in the official rankings LFP 2022–2023 (up) and when using the rankings computed by the proposed method (down)

3.2 LFP Season 2023–2024

As before, the starting point of the proposed method is the ranking r_0 corresponding to the previous season, replacing the descended teams by the promoted ones. The ranking r_0 to be used as starting point for the season 2023–2024 was already shown in Table 2.

In Table 11 we show the classifications produced with the proposed method along the last six weeks. If we focus on the evolution of the Top 6 teams, it is easy to note the great difference between these rankings and the official rankings that were given in Table 1. For example in Table 1 only seven teams appear in the Top 6 positions, while in Table 11 a total of 17 teams contribute to the Top 6 ranking (the teams that do not appear being Cádiz, Valencia and Las Palmas). A quantified analysis of the differences in the classifications produced by both methods is presented in the subsequent sections.

Table 10 General classification for season 2022–2023 as given by the new method and by the official system

Position	New method	Points	Official LFP	Points
1	Barcelona	411	Barcelona	88
2	Real Madrid	376	Real Madrid	78
3	Atlético	364	Atlético	77
4	R. Sociedad	285	R. Sociedad	71
5	Villarreal	282	Villarreal	64
6	Betis	191	Betis	60
7	Athletic	187	Osasuna	53
8	Girona	178	Athletic	51
9	Sevilla	171	Mallorca	50
10	Osasuna	159	Girona	49
11	Almería	159	Rayo	49
12	Valencia	154	Sevilla	49
13	Mallorca	149	Celta	43
14	Celta	144	Cádiz	42
15	Rayo	140	Getafe	42
16	Real Valladolid	115	Valencia	42
17	Getafe	109	Almería	41
18	Cádiz	109	Real Valladolid	40
19	Espanyol	92	Espanyol	37
20	Elche	63	Elche	25

3.2.1 Comparison of Season Competitiveness as Measured by NS

By using the measure NS we can assign a competitiveness index to a season. In Table 12 we show the values of NS corresponding to the series of rankings produced by the proposed method and by using the official rankings for season 2023–2024. As a conclusion from Table 12 we observe that the value of NS given by the novel method is much bigger than the one corresponding to the series given by the official LFP system. Therefore, our method gives more CB to the championship than the official method, as occurred in the previous season.

3.2.2 Comparative Analysis of Teams Obtaining Victories in Individual Rounds

In Fig. 6 we show the teams that attain the top ranking in the official LFP standings for the 2023–2024 season, alongside those that emerge as leaders when employing the new methodology for the same seasonal period. We observe that only four teams

Table 11 Standings computed with the proposed method for the last six weeks of LFP 2023–2024

W. 33	W. 34	W. 35	W. 36	W. 37	W. 38
Alavés	Real Madrid	Real Madrid	Real Madrid	Osasuna	Girona
Villarreal	Sevilla	Barcelona	Atlético	Barcelona	Almería
Granada	Girona	Betis	Barcelona	Girona	Atlético
Barcelona	Betis	Villarreal	Rayo	R. Sociedad	Barcelona
Getafe	R. Sociedad	Atlético	Celta	Athletic	Mallorca
Atlético	Athletic	Mallorca	Villarreal	Celta	Athletic
Girona	Celta	Cádiz	Cádiz	Alavés	Celta
Real Madrid	Alavés	Girona	R. Sociedad	Real Madrid	Valencia
Sevilla	Atlético	Athletic	Betis	Villarreal	Osasuna
Betis	Almería	Alavés	Las Palmas	Mallorca	Alavés
Mallorca	Villarreal	Osasuna	Mallorca	Almería	Villarreal
Cádiz	Mallorca	Valencia	Osasuna	Cádiz	Las Palmas
R. Sociedad	Valencia	Rayo	Athletic	Las Palmas	Real Madrid
Valencia	Rayo	Sevilla	Granada	Granada	Betis
Athletic	Barcelona	Almería	Girona	Getafe	Getafe
Almería	Getafe	Celta	Valencia	Valencia	Sevilla
Las Palmas	Las Palmas	Getafe	Sevilla	Betis	Rayo
Celta	Osasuna	Las Palmas	Almería	Sevilla	R. Sociedad
Rayo	Granada	R. Sociedad	Getafe	Atlético	Cádiz
Osasuna	Cádiz	Granada	Alavés	Rayo	Granada

Table 12 Values of NS

Season	LFP	New method
2023–2024	0.047	0.433

accessed to the first place with the official method, while 12 teams accessed the first position with the novel proposal. That is, our method gives more chance to the teams to win a round.

3.2.3 Comparison of Runner-Up Teams in Individual Rounds

In Fig. 7 we show the teams that occupy the second position in the official LFP rankings for the season 2023–2024, and the teams that occupy the same position by using the proposed method for the same seasonal period. Note that only five teams achieved the second place with the official method, while 15 teams accessed the same position with the novel proposal. Therefore, our method gives more chance to attain a second position in an individual round.

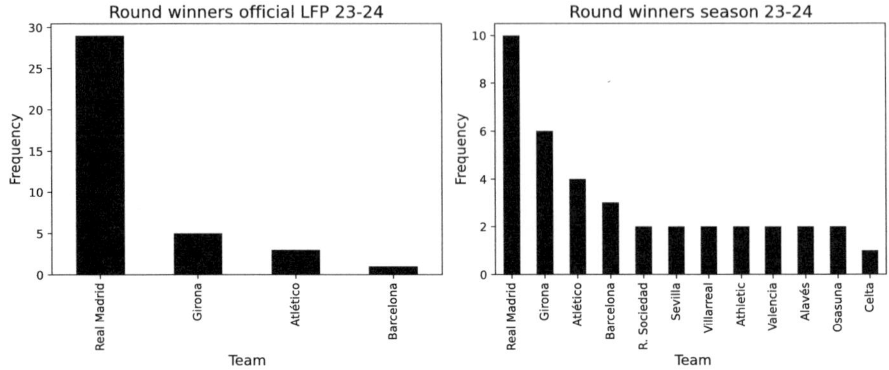

Fig. 6 Teams occupying the first position at any round in the official rankings (left panel) and by the proposed method, in season LFP 2023–2024

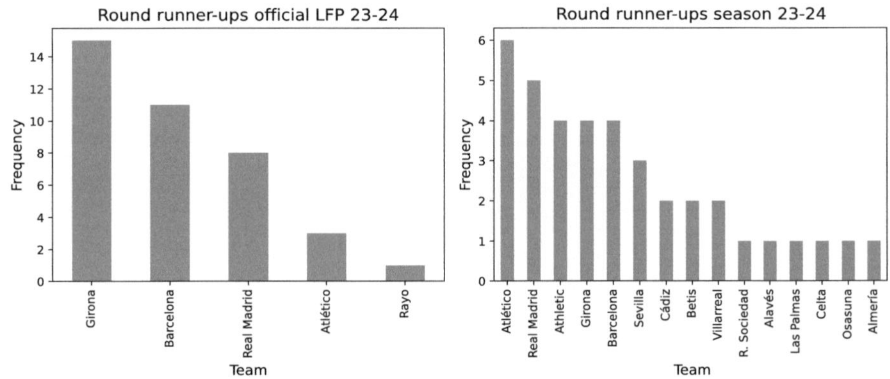

Fig. 7 Teams that occupy the second position at any round in the official (left panel) and by the proposed method, in season LFP 2023–2024

3.2.4 Comparison of Teams Occupying the Third Position in Individual Rounds

In Fig. 8 we show the teams occupying the third position in the official LFP rankings for the season 2023–2024 (left panel), and the teams that occupy the same position when using the rankings computed by the new method. Note that only six teams achieved the third place with the official method, while 17 teams accessed the same position with the new proposal.

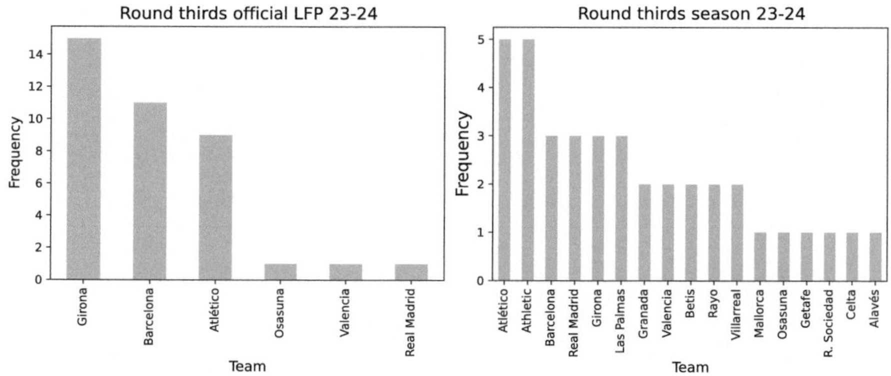

Fig. 8 Teams occupying the third position in the official (left panel) and by the proposed method, in season LFP 2023–2024

3.2.5 Comparative Examination of Teams Occupying the Top 3 Positions

As we have shown in Table 1 the Top 3 teams for the official classification of LFP 2023–2024 are Real Madrid, FC Barcelona, and Girona FC. In Fig. 9 (upper panel) we show the times that these teams occupy the first, second and third positions in the official standings. As we see, the FC Barcelona occupied the first position during 28 weeks, while the Real Madrid occupied the first position only 8 times. In Fig. 9 (lower panel) we show the teams that occupied simultaneously the first Top 3 positions along the championship. As happened with the previous season we obtain that the novel method offers more chance to attain one of the three top places.

Another form to compare the movement of the teams along the rankings is by focusing only on the Top 3 teams. We do this in Fig. 10 where we compare the movements along the classifications of Real Madrid, FC Barcelona, and Girona FC when using the official classifications (upper panel) and when using the rankings computed by the proposed method (lower panel). It is clear that the new method shows a more random behaviour that can be interpreted as an attractive factor for fans engagement.

3.2.6 General Classification for LFP 2023–2024

The general classification derived from our proposed methodology for the 2023–2024 LFP season is presented in Table 13, alongside the official classification for that season. Several notable features emerge from this table. Firstly, the teams qualifying for the Champions League competition are identical in both classifications, albeit with differing rankings, since they comprise the top four teams. Secondly, the teams qualifying for the Europa League are also identical in both systems, namely the fifth and sixth ranked teams. As observed in the previous season, the systems

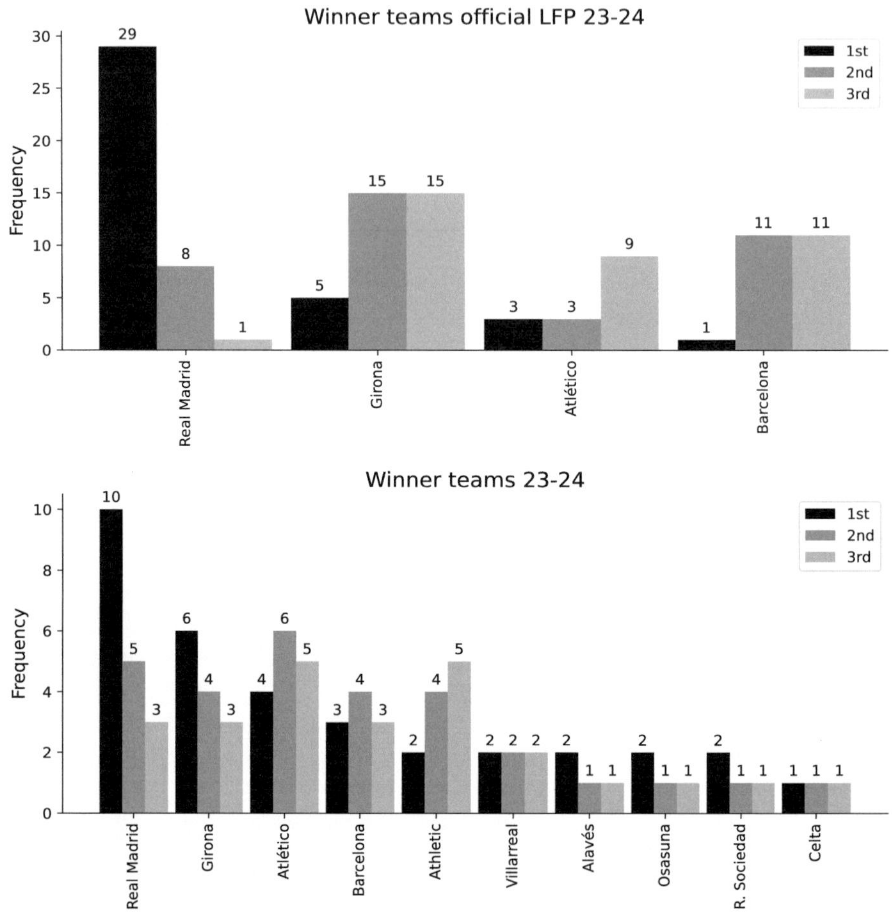

Fig. 9 Times that the Top 3 teams achieved the first, second and third place in the official rankings LFP 2023–2024 (up) and when using the rankings computed by the proposed method (down)

diverge regarding the team qualified for the Conference League championship, which occupies the seventh rank. Finally, a further notable feature is that the systems concur on the teams relegated to lower divisions, comprising the bottom three teams. We consider these similarities with the official system as evidence supporting the consistency of our novel methodology.

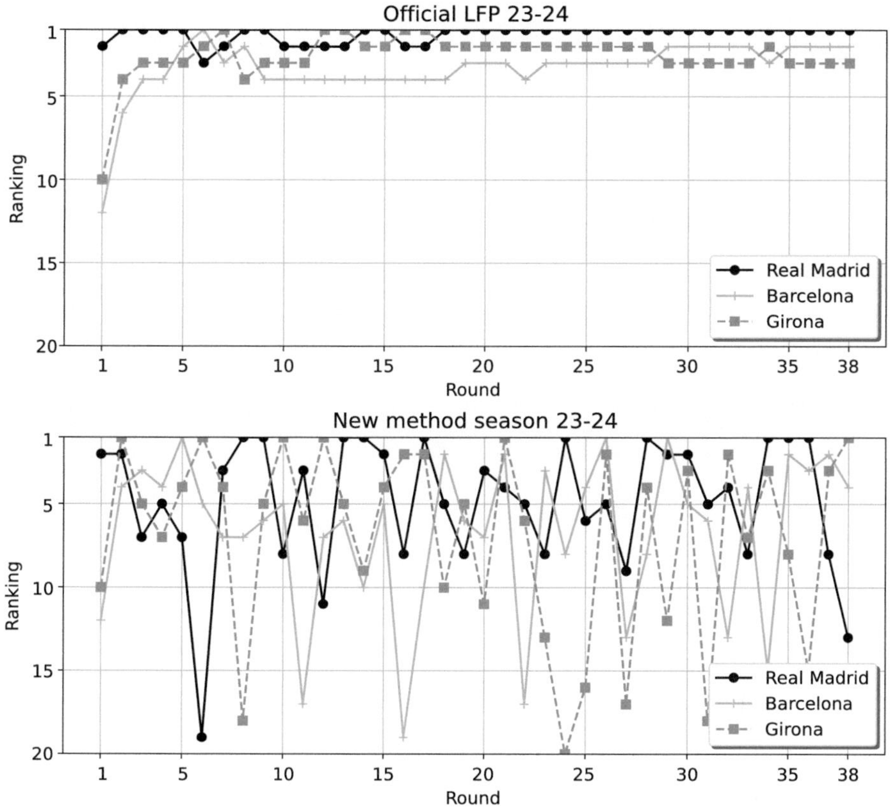

Fig. 10 Movements along the classifications of Real Madrid, FC Barcelona, and Girona FC in the official rankings LFP 2023–2024 (up) and when using the rankings computed by the proposed method (down)

4 Conclusions

We have proposed a novel approach for ranking teams on a weekly basis throughout the first division of the Spanish La Liga Primera División (LFP) championship. Furthermore, we develop a method for obtaining an overall classification by sharing points, inspired by the Formula 1 championships. The proposed methodology is applied to the 2022–2023 and 2023–2024 seasons of the LFP, yielding results that demonstrate the new approach produces more pronounced movements in the rankings. This is reflected in the increased value of the competitive balance coefficient NS for both seasons (cf. Sects. 3.1.1 and 3.2.1). Notably, the novel method yields a final ranking that closely resembles the official final ranking for the studied seasons when considering specific table positions. In particular, our findings highlight the following conclusions:

Table 13 General classification for season 2023–2024 as given by the new method and by the official system

Position	New method	Points	Official LFP	Points
1	Real Madrid	505	Real Madrid	95
2	Girona	391	Barcelona	85
3	Atlético	369	Girona	81
4	Barcelona	358	Atlético	76
5	Athletic	296	Athletic	68
6	R. Sociedad	210	R. Sociedad	60
7	Villarreal	193	Betis	57
8	Valencia	167	Villarreal	53
9	Osasuna	162	Valencia	49
10	Betis	162	Alavés	46
11	Sevilla	153	Osasuna	45
12	Las Palmas	143	Getafe	43
13	Celta	134	Celta	41
14	Alavés	131	Sevilla	41
15	Getafe	118	Mallorca	40
16	Rayo	98	Las Palmas	40
17	Mallorca	73	Rayo	38
18	Granada	72	Cádiz	33
19	Cádiz	65	Almería	21
20	Almería	38	Granada	21

1. As observed in Sects. 3.1.2 and 3.2.2, our approach yields a higher number of teams attaining the top position at any given round, thereby introducing the possibility of establishing a *Champion of the Round* title, potentially sponsored by a company. This innovation is likely to increase fan interest. Furthermore, as demonstrated in Sects. 3.1.3, 3.1.4, 3.2.3, and 3.2.4, the novel method offers teams more opportunities to secure one of the top three positions, primarily due to its emphasis on current-round results.
2. As deduced from Sects. 3.1.5 and 3.2.5, when examining the movements of Top 3 teams, it becomes apparent that our proposed method exhibits a more random behavior, which can be interpreted as an attractive factor for fan engagement.
3. The results presented in Sects. 3.1.6 and 3.2.6 suggest that the similarities observed with respect to teams qualifying for the UEFA Champions League, the UEFA Conference League, and relegated teams under our novel method serve as an indication to its consistency.

Future research directions are being pursued along two lines. Firstly, a real-time computational framework for classification is being developed, with the ultimate goal of facilitating instantaneous updates throughout the current season. However, a key challenge arises from unforeseen alterations to the scheduled fixtures (e.g., postponed matches due to various causes), necessitating the provision of interim classifications pending the resolution of adjourned encounters. Secondly, an examination of additional tie-breaking rules is underway, aimed at further enriching our proposed methodology.

Competing Interests The authors have no conflicts of interest to declare that are relevant to the content of this chapter.

References

1. Avila-Cano, A., Triguero-Ruíz, F.: On the control of competitive balance in the major European football leagues. Manag. Decis. Econ. **44**(2), 1254–1263 (2022)
2. Barrat, A., Barthélemy, M., Pastor-Satorras, R., Vespignani, A.: The architecture of complex weighted networks. Proc. Natl. Acad. Sci. **101**(11), 3747–3752 (2004)
3. Biedl, T., Brandenburg, F.J., Deng, X.: Crossings and permutations. In: Healy, P., Nikolov, N.S. (eds.) LNCS, vol. 3843, pp. 1–12. Springer, Berlin, Heidelberg (2005)
4. Criado, R., García, E., Pedroche, F., Romance, M.: A new method for comparing rankings through complex networks: Model and analysis of competitiveness of major European soccer leagues. Chaos Interdiscip. J. Nonlinear Sci. **23**, 043114 (2013)
5. Criado, R., García, E., Pedroche, F., Romance, M.: On graphs associated to sets of rankings. J. Comput. Appl. Math. **291**, 497–508 (2016)
6. Csató, L.: How to avoid uncompetitive games? The importance of tie-breaking rules. Eur. J. Oper. Res. **307**, 1260–1269 (2023)
7. Diaconis, P., Graham, R.L.: Spearman's footrule as a measure of disarray. J. Roy. Stat. Soc. B. Met. **39**(2), 262–268 (1977)
8. Dobson, S., Goddard, J.: The Economics of Football, 2nd edn. Cambridge University Press (2011)
9. Emond, E.J., Mason, D.W.: A new rank correlation coefficient with application to the consensus ranking problem. J. Multi-Crit. Decis. Anal. **11**, 17–28 (2002)
10. Fort, R., Maxcy, J.: Competitive balance in sports leagues: an introduction. J. Sport. Econ. **4**, 154–160 (2003)
11. Goossens, D.R., Beliën, J., Spieksma, F.C.R.: Comparing league formats with respect to match importance in Belgian football. Ann. Oper. Res. **194**, 223–240 (2012)
12. Groot, L.: Economics, Uncertainty and European FootballâŁ¯: Trends in Competitive Balance. Edward Elgar Cop, Northampton, USA (2008)
13. Kendall, M.G.: A new measure of rank correlation. Biometrika **30**, 81–89 (1938)
14. Kendall, M.G.: Rank Correlation Methods, 4th edn. Griffin, London, UK (1970)
15. Kendall, M.G., Babington-Smith, B.: The problem of m rankings. Ann. Math. Stat. **10**, 275–287 (1939)
16. Kesenne, S.: How can the competitive balance be improved? In: García, J. (ed.) Sports (and) Economics, p. 57. Funcas. Madrid (2019)
17. Pedroche, F.: Competitiveness of Formula 1 championship from 2012 to 2022 as measured by Kendall corrected evolutive coefficient. In: Torregrosa, J.R., et al. (ed.) Mathematical Modelling in Engineering & Human Behaviour 2023, pp. 394–404 (2023)
18. Pedroche, F., Conejero, J.A.: Corrected evolutive Kendall's τ coefficients for incomplete rankings with ties: application to case of spotify lists. Mathematics **8**(10), 1828 (2020)
19. Pedroche, F., Verdoy, J.A.: Comparison of some measures of competitive balance in regular seasons of European basketball leagues. In: Jodar, L., et al. (ed.) Mathematical Modelling in Engineering & Human Behaviour 2014, pp. 106–110 (2014)

20. Rottenberg, S.: The Baseball player's labor marquet. J. Polit. Econ. **64**(3), 242–258 (1956)
21. https://resultados.rfef.es/pnfg/NPortada
22. Real Federación Española de Fútbol. Normas Reguladoras y Bases de Competición de Primera y Segunda División de Fútbol. Temporada 2023/2024
23. https://www.laliga.com/en-ES/transparency/institutional-information

Simultaneous Recovery of a Sparse Topology and the Admittance of an Electrical Network

Álvaro Samperio

Abstract We show that the problem of recovering the topology and admittance of an electrical network from power and voltage data at all vertices is often ill-posed, and sometimes it even has multiple solutions. We reformulate the problem to seek for a sparse network, i.e., with few edges, which fits the data up to a given tolerance. We propose an algorithm to solve this reformulated problem. It combines, in an iterative procedure, the resolution of non-negative linear regression problems, and techniques of spectral graph sparsification. The algorithm is based on original results bounding the fitting error of a sparse approximation of a network. We illustrate our techniques with several experimental results in which we are able to recover a sparse network.

Keywords Graph sparsification · Ill-posed inverse problems · Electrical networks · Sparse linear regression

1 Introduction and Preliminaries

We begin with some notations, definitions and results of vector calculus on electrical networks from [12]. Given $m \in \mathbb{N}^* = \mathbb{N} \setminus \{0\}$ and a finite set V, we denote the spaces of real and complex vector functions on V as $\mathscr{C}(V, \mathbb{R}^m) = \{u\colon V \longrightarrow \mathbb{R}^m\}$ and $\mathscr{C}(V, \mathbb{C}^m) = \{u\colon V \longrightarrow \mathbb{C}^m\}$, respectively. In the particular case of real scalar functions, we denote $\mathscr{C}(V) = \mathscr{C}(V, \mathbb{R})$. We denote the set of nonnegative scalar functions on V as $\mathscr{C}^+(V)$. Given $\boldsymbol{u} \in \mathscr{C}(V, \mathbb{C}^m)$, respectively $\boldsymbol{u} \in \mathscr{C}(V, \mathbb{R}^m)$, the components of $\boldsymbol{u}$ are $u_1, \dots, u_m \in \mathscr{C}(V, \mathbb{C})$, respectively $u_1, \dots, u_m \in \mathscr{C}(V)$, i.e. $\boldsymbol{u} = (u_1, \dots, u_m)$. We denote its real and imaginary parts as $\Re(\boldsymbol{u}) = (\Re(u_1), ..., \Re(u_m)) \in \mathscr{C}(V, \mathbb{R}^m)$ and $\Im(\boldsymbol{u}) = (\Im(u_1), ..., \Im(u_m)) \in \mathscr{C}(V, \mathbb{R}^m)$, and its complex conjugate as $\overline{\boldsymbol{u}} = (\overline{u_1}, \dots, \overline{u_m})$.

Given $F \subseteq V$, we denote by $\mathscr{C}(F)$ the subspace of the functions in $\mathscr{C}(V)$ which are null on $V \setminus F = F^c$. Also, for each $u \in \mathscr{C}(V, \mathbb{C})$ we define $\int_F u dx = \sum_{x \in F} u(x)$. On $\mathscr{C}(V, \mathbb{C}^m)$ we consider the inner product

Á. Samperio (✉)
Departamento de Matemáticas, CUNEF Universidad, Madrid, Spain
e-mail: alvaro.samperio@cunef.edu

J. R. Torregrosa (ed.), *Mathematical Modelling in Engineering Human Behaviour*, Springer Proceedings in Mathematics & Statistics 513,
https://doi.org/10.1007/978-3-032-00567-0_4

$$\langle \boldsymbol{u}, \boldsymbol{v} \rangle = \sum_{j=1}^{m} \langle u_j, v_j \rangle = \sum_{j=1}^{m} \int_V u_j \bar{v}_j dx,$$

whose associated norm is denoted by $||\cdot||_2$, or just by $||\cdot||$. This inner product induces an inner product on $\mathscr{C}(V, \mathbb{R}^m)$. We also consider in $\mathscr{C}(V, \mathbb{C}^m)$ the norm $||\cdot||_\infty$, which is defined for each $\boldsymbol{u} = (u_1, ..., u_m) \in \mathscr{C}(V, \mathbb{C}^m)$ as $||\boldsymbol{u}||_\infty = \max_{j=1,..,m} \left\{||u_j||_\infty\right\}$.

A Direct Current (DC) network is modeled mathematically as a graph with a real positive weight at each edge, which is the inverse of its *resistance*. For modeling Alternating Current (AC) networks there are different alternatives, depending on the characteristics of the network. In this work, we consider balanced three-phase networks in which all lines are inductive and "short", (i.e., their length is shorter than 80 km). In that case, the *susceptance to earth* of the lines can be neglected and the mathematical model of a network is a graph in which each physical connection between a pair of vertices is represented by a single edge characterized by a complex weight with nonnegative real part and nonpositive imaginary part called *admittance*, (see [17, 18]).

Definition 1 An *electrical network* is a pair $\Gamma = (V, a)$ where V is a finite nonempty set called *vertex set*, and $a \in \mathscr{C}(V \times V, \mathbb{C})$, called *admittance*, satisfies the following properties:

(i) $a = c - ib$ where $c, b \in \mathscr{C}^+(V \times V)$ and the function c is called *conductance* whereas the function b is called *susceptance*.
(ii) c and b are symmetric and $c(x, x) = b(x, x) = 0$, for any $x \in V$.

An electrical network $\Gamma = (V, a)$ has an associated graph $G(\Gamma) = (V, E(\Gamma))$ called its "*network topology*". Its vertex set is V and its *edge set* is $E(\Gamma) = \big\{\{x, y\}$ such that $x, y \in V$ and $a(x, y) \neq 0\big\}$. We say that $x, y \in V$ are *adjacent* iff $\{x, y\} \in E(\Gamma)$, and we denote $e_{xy} = e_{yx} = \{x, y\}$. For any $e_{xy} \in E(\Gamma)$, we denote by $a(e_{xy})$ the value $a(x, y) = a(y, x) \neq 0$. Note that $a(x, x) = 0$ for any $x \in V$, so $(V, E(\Gamma))$ is a finite undirected simple graph without loops, i.e., without edges from a vertex to itself.

In the particular case the network is operating in DC, $b = 0$, so the network $\Gamma = (V, c)$ is a real weighted graph.

For an AC network $\Gamma = (V, a)$ we consider the two DC networks $\Gamma_c = (V, c)$ and $\Gamma_b = (V, b)$, which we will call the conductance and susceptance networks of Γ, respectively. Note that $E(\Gamma) = E(\Gamma_c) \cup E(\Gamma_b)$.

For any function $a \in \mathscr{C}(V \times V, \mathbb{C})$ that is symmetric and satisfies that $a(x, x) = 0$ for any $x \in V$, we denote by $\mathcal{L}_a$ the endomorphism of $\mathscr{C}(V, \mathbb{C})$ defined as

$$\mathcal{L}_a(u)(x) = \sum_{y \in V} a(x, y) \big(u(x) - u(y)\big)$$

for any $u \in \mathscr{C}(V, \mathbb{C})$ and any $x \in V$. When a is the admittance of an electrical network $\Gamma = (V, a)$, $\mathcal{L}_a$ is called the *Laplacian* of Γ. *Gauss' Theorem* states that for any $u \in \mathscr{C}(V, \mathbb{C})$, $\int_V \mathcal{L}_a(u)dx = 0$.

If $\Gamma = (V, a)$ is an AC network, at a given time, the potential at the vertices can be represented by $u \in \mathscr{C}(V, \mathbb{C})$, the current injected at the vertices is equal to $\mathcal{L}_a(u) \in \mathscr{C}(V, \mathbb{C})$ and the power injected at the vertices is equal to $s = u\overline{\mathcal{L}_a(u)} \in \mathscr{C}(V, \mathbb{C})$. Similarly, if $\Gamma = (V, c)$ is a DC network, the potential at the vertices can be represented by $u \in \mathscr{C}(V)$, the current injected at the vertices is equal to $\mathcal{L}_c(u) \in \mathscr{C}(V)$, and the power injected at the vertices is equal to $s = u\mathcal{L}_c(u) \in \mathscr{C}(V)$. These equations relating voltage and power are called the *power flow equations*, (see [11]).

Remark 1 In order to avoid possible faults in the operation of an electrical network, voltages are usually maintained close to a reference value, that is chosen to be $u = 1$ in appropriate units (see [11]). In this chapter we will fix minimum and maximum values of voltage $0 < |u|_{\min} \leq |u|_{\max}$. Later, in the experiments, we will use $|u|_{\min} = 0.9$ and $|u|_{\max} = 1.1$.

Let $\Gamma = (V, a)$ be an electrical network and let $F \subsetneq V$. We say that a function u is harmonic on F when $\mathcal{L}(u) = 0$ on F. Given $g \in \mathscr{C}(F^c, \mathbb{C})$, the *Dirichlet problem* of finding $u \in \mathscr{C}(V, \mathbb{C})$ such that

$$\mathcal{L}_a(u) = 0 \quad \text{on} \quad F, \qquad u = g \quad \text{on} \quad F^c, \tag{1}$$

always has a solution. Moreover, the images of all the solutions to (1) by the Laplacian $\mathcal{L}_a$ are equal. Furthermore, there is an endomorphism Λ of $\mathscr{C}(F^c, \mathbb{C})$ called the *Dirichlet-to-Neumann map of* Γ *and* F defined for each $g \in \mathscr{C}(F^c, \mathbb{C})$ as $\Lambda(g) = \mathcal{L}_a(u_g)$, where u_g is a solution to (1), (see [12]).

That is, the Dirichlet-to-Neumann map is the linear map which gives us the relationship between the potential at F^c and its corresponding current injected at F^c under the condition that the potential is harmonic on F.

Additionally, for all DC networks and for some AC networks, the Dirichlet-to-Neumann map of Γ and F is the Laplacian of another network whose vertex set is F^c and whose admittance we denote by a^Λ, which is called the *Kron reduction of* Γ *with respect to* F. When x and y are two distinct vertices of V, the Dirichlet-to-Neumann map of Γ and $F = V \setminus \{x, y\}$ is always a Kron reduction (with two vertices), and the value $a^e(x, y) := a^\Lambda(x, y)$ is called the *effective admittance between* x *and* y. When Γ is a DC network, the effective admittance is nonnegative, it is called effective conductance and it is denoted as $c^e(x, y)$.

Denoting by $c^e_{\setminus e_{xy}}(x, y)$ the effective conductance between x and y in the network obtained from a DC network Γ by removing the edge e_{xy}, we have that, (see [12]),

$$c^e(x, y) = c^e_{\setminus e_{xy}}(x, y) + c(x, y), \tag{2}$$

and thus $c^e(x, y) \geq c(x, y)$. For every edge e_{xy} we define the *effective resistance between* x *and* y as $r^e(x, y) = 1/c^e(x, y) > 0$.

We define the *energy* of a DC network $\Gamma = (V, c)$, (see [4]), as the positive semidefinite symmetric bilinear form on $\mathscr{C}(V)$ given by

$$\mathcal{E}(u, v) = \langle u, \mathcal{L}_c(v) \rangle = \int_V u \mathcal{L}_c(v) dx.$$

Nowadays, electrical systems are expanding and increasing in complexity really fast due to the deregulation and proliferation of distributed energy resources. Because of this expansion, sometimes the system operators do not have precise and updated information about the admittance or the topology of the network. Also, the dependence of admittance on temperature can cause the information about it to be incorrect [1]. The lack of topology information is especially frequent in the secondary distribution network [10], in which the lines are short [3]. Moreover, recently, the availability and accuracy of measured data in the electrical system has increased significantly. At the nodes, power injection and voltage magnitude can be measured using smart meters and the phase angle information can be obtained by micro phasor measurement units [10]. In that scenario, the simultaneous recovery of the topology and admittance of a network from voltage and power data is of great interest from an applied point of view.

Several authors have formulated different inverse problems on networks which involve the recovery of the topology, starting from different data; and they have proposed methods to solve these problems. For instance, in the context of the inverse conductance problem, in [9] the authors introduced a method which recovers the conductance and the topology up to vertices of combinatorial degree two of any DC tree from its Dirichlet-to-Neumann map.

In [1], the application of a least squares approach to solve two problems is discussed: the first one concerns the admittance estimation alongside voltage estimation from a known topology, and the second one consists of the voltage estimation alongside detection of errors in the network topology determined by a given admittance. However, in that paper, the simultaneous recovery of topology and admittance is not discussed.

An iterative method is proposed in [10] to recover simultaneously the topology and admittance of a network from power and voltage data, starting at each step from an estimated admittance and then calculating the current flow in each edge, removing an edge if its conductance and susceptance have a value below a given threshold. Nevertheless, relying only on the values of the admittance to decide if an edge can be removed from a network can lead us to an error, as we show in Example 3. The method proposed in [6] also estimates both the topology and admittance of a network from power and voltage data, but only in the particular case of distribution networks which operationally are a tree.

In this work, we study the problem of simultaneously recovering the topology and admittance of an electrical network from power and voltage data at all vertices. We introduce the following notation to formulate the general network recovery problem with exact data (Problem 1).

For any $a \in \mathscr{C}(V \times V, \mathbb{C})$ that is symmetric and satisfies that $a(x, x) = 0$ for any $x \in V$, we denote by $\mathscr{P}_a : \mathscr{C}(V, \mathbb{C}^m) \times \mathscr{C}(V, \mathbb{C}^m) \longrightarrow \mathscr{C}(V, \mathbb{C}^m)$ the function defined for each $(\boldsymbol{v}, \boldsymbol{w}) = ((v_1, ..., v_m), (w_1, ..., w_m)) \in \mathscr{C}(V, \mathbb{C}^m) \times \mathscr{C}(V, \mathbb{C}^m)$ as

$$\mathscr{P}_a (\boldsymbol{v}, \boldsymbol{w}) = (v_1 \overline{\mathcal{L}_a(w_1)}, ..., v_m \overline{\mathcal{L}_a(w_m)}).$$

Of course, when a is a real function, $\mathscr{P}_a (\boldsymbol{v}, \boldsymbol{w}) \in \mathscr{C}(V, \mathbb{R}^m)$ for every $\boldsymbol{u}, \boldsymbol{v} \in \mathscr{C}(V, \mathbb{R}^m)$. Also, if a is the admittance of an AC network, respectively a DC network, then for any $\boldsymbol{u} \in \mathscr{C}(V, \mathbb{C}^m)$, respectively $\boldsymbol{u} \in \mathscr{C}(V, \mathbb{R}^m)$, the j-th component of the function $\mathscr{P}_a (\boldsymbol{u}, \boldsymbol{u})$ is equal to the power injected at the network when the potential is u_j, for any $j = 1, ..., m$.

We denote as $E(K_V)$ the set of edges of the complete graph on V, i.e. the set of all possible edges on the vertex set V. In an applied situation, one of the main causes that the system operators of a network Γ may not know its topology $(V, E(\Gamma))$ is the existence of lines which have a switch whose status is unknown, so they do not know whether their corresponding edges belong to $E(\Gamma)$ or not, (see [1]). In general, we formulate the inverse problem of recovering the topology and admittance of Γ with the a priori information that there is a set $E \subseteq E(K_V)$ of edges which are candidates to belong to $E(\Gamma)$, and we know that the edges in $E(K_V) \setminus E$ do not exist in the network topology. Note that, the particular case in which $E = E(K_V)$, corresponds with the framework in which we do not have any a priori information about $E(\Gamma)$.

Problem 1 *Let V be a finite set of vertices, let E be a set of edges on V, let $0 < |u|_{\min} \leq |u|_{\max}$ be positive values, let $m \in \mathbb{N}^*$ and let $\boldsymbol{u}, \boldsymbol{s} \in \mathscr{C}(V, \mathbb{C}^m)$, respectively $\boldsymbol{u}, \boldsymbol{s} \in \mathscr{C}(V, \mathbb{R}^m)$, such that $|u|_{\min} \leq |u_j| \leq |u|_{\max}$ for all $j = 1, \ldots, m$. Determine an AC, respectively a DC, electrical network $\Gamma = (V, a)$ with set of edges $E(\Gamma) \subseteq E$ such that $\boldsymbol{s} = \mathscr{P}_a(\boldsymbol{u}, \boldsymbol{u})$.*

The pair $(\boldsymbol{u}, \boldsymbol{s})$ is usually called *the data pair* or *the data set* and hence we always understand that for any $j = 1, \ldots, m$, u_j represents a measurement of voltage and s_j represents its corresponding measurement of power. Therefore for any $j = 1, \ldots, m$ we have that $u_j(z) \neq 0$ for all $z \in V$, since $0 < |u|_{\min} \leq |u_j|$. Note that the set of admittances, respectively conductances, on $V \times V$ that define a network Γ satisfying $E(\Gamma) \subseteq E$ can be identified with $\mathscr{C}^+(E) \times \mathscr{C}^+(E)$, respectively $\mathscr{C}^+(E)$.

In Sect. 2 we show that Problem 1 is ill-posed. Therefore, we usually obtain a solution whose set of edges is the whole E. If E is large, the recovered network is inefficient for solving usual applied problems in electrical networks which require knowing the topology and admittance. Those applications include failure identification, power flow optimization or generation scheduling [6]. In Sect. 3 we reformulate the problem to look for a sparse network whose fitting error to the data is lower than a fixed tolerance, (see Problem 3). Recovering a network with those characteristics would allow the efficient and accurate resolution of those applied problems.

In Sect. 4 we review the notion of spectral sparsification of a network introduced in [16] by Spielman and Teng. We also review and give a novel physical interpretation to Algorithm 1, which was proposed in [2] to construct a sparse spectral approximation of a network by removing edges from it. In Sect. 5, we prove original theoretical results (see Theorems 2 and 3) that give an upper bound on the increase of the fitting error of a network in that process of removing edges.

These theoretical results are the basis for the Algorithm 2 that we propose in Sect. 6 to solve Problem 3. The algorithm consists in an iterative procedure of recovering a network and sparsifying it, with an automatic exploration to find the levels of closeness of sparsification that allow us to remove edges without increasing the error in the data above the fixed tolerance.

Finally, in Sect. 7 we show experimental results of the application of Algorithm 2. We see that, in some cases, we recover a network with the actual topology and low error in the admittance. We show other cases in which we recover a Kron reduction of the network alongside isolated nodes when the power injected at them is always equal to zero in the data set. Additionally, we show that, in some cases, the algorithm has the advantage of providing a sparse approximation of the real network, in which the computation of the mentioned applied problems is faster than in the real network.

2 Ill-Posedness of the Network Recovery Problem

Problem 1 is ill-posed in general, often having multiple solutions. For example, in the following example, we show a case in which, under the restriction that the power injected at a node is zero at a certain vertex for all the s_j functions in the data pair $(\boldsymbol{u}, \boldsymbol{s})$, then there are multiple solutions for any choice of $(\boldsymbol{u}, \boldsymbol{s})$. This restriction is common in networks of the real world, in which there are vertices which are not associated to any generator or consumer of power.

Example 1 Let $\Gamma = (V, c)$ be the DC electrical network which is the weighted path in Fig. 1, with vertex set $V = \{x, y, z\}$, with set of edges $E(\Gamma) = \left\{e_{xy}, e_{yz}\right\}$, whose conductance values are $c(x, y) = c(y, x) > 0$ and $c(y, z) = c(z, y) > 0$, respectively; and with Laplacian $\mathcal{L}$. Let $(\boldsymbol{u}, \boldsymbol{s})$ be a data pair satisfying the power flow equations, that is, $\boldsymbol{s} = \mathscr{P}_c(\boldsymbol{u}, \boldsymbol{u}) = \left(u_1\mathcal{L}(u_1), \ldots, u_m\mathcal{L}(u_m)\right)$, and moreover that $\boldsymbol{s}(y) = \boldsymbol{0}$.

We have that for each $j = 1, ..., m$, u_j is harmonic on y, because $u_j(y) \neq 0$. By Gauss' Theorem, $\mathcal{L}(u_j)(x) = I_j$, $\mathcal{L}(u_j)(y) = 0$ and $\mathcal{L}(u_j)(z) = -I_j$, for some $I_j \in \mathbb{R}$. Then, each u_j must be equal to $u_j(x) = \frac{I_j}{c(x,y)} + \zeta_j$, $u_j(y) = \zeta_j$ and $u_j(z) = -\frac{I_j}{c(y,z)} + \zeta_j$, for some $\zeta_j \in \mathbb{R}$.

We consider the Problem 1 with data pair $(\boldsymbol{u}, \boldsymbol{s})$ and the edge set of the complete graph in V, $E(K_V) = \left\{e_{xy}, e_{xz}, e_{yz}\right\}$. Then, a DC electrical network $\Gamma' = (V, c')$

Fig. 1 Electrical network in Example 1

with Laplacian $\mathcal{L}'$ is a solution of the problem if and only if we have that $c' \in \mathscr{C}^+(E)$ is a solution of

$$\begin{cases} c'(x,y)\big(u_j(x)-u_j(y)\big)+c'(x,z)\big(u_j(x)-u_j(z)\big)=I_j \\ c'(x,z)\big(u_j(z)-u_j(x)\big)+c'(y,z)\big(u_j(z)-u_j(y)\big)=-I_j \\ c'(x,y)\big(u_j(y)-u_j(x)\big)+c'(y,z)\big(u_j(y)-u_j(z)\big)=0, \end{cases}$$

for all $j=1,...,m$. For each $j=1,...,m$, the third equation in that system depends linearly on the rest. If $I_j=0$, then u_j is constant on V, so any value of the conductance c' satisfies that $s_j=u_j\mathcal{L}'(u_j)$. If, on the contrary, $I_j\neq 0$, then the previous system is equivalent to the following system, which is independent of I_j and ζ_j:

$$\begin{cases} \dfrac{c'(x,y)}{c(x,y)}+\left(\dfrac{1}{c(x,y)}+\dfrac{1}{c(x,z)}\right)c'(x,z)=1 \\ \left(\dfrac{1}{c(x,y)}+\dfrac{1}{c(x,z)}\right)c'(x,z)+\dfrac{c'(y,z)}{c(x,z)}=1. \end{cases} \tag{3}$$

The function $\alpha \in \mathscr{C}(E)$; which is determined by the values $\alpha(x,y)=-1-\frac{c(x,y)}{c(y,z)}$, $\alpha(x,z)=1$ and $\alpha(y,z)=-1-\frac{c(y,z)}{c(x,y)}$; is a solution to the homogeneous system associated to (3). Therefore, any network $\Gamma'=(V,c')$ whose conductance is of the form $c'=c+\delta\alpha$, with $\delta\in\left[0,\frac{c(x,y)c(y,z)}{c(x,y)+c(y,z)}\right]$, is a solution to Problem 1.

Among the solutions with conductance of this form $c'=c+\delta\alpha$, the one with $\delta=0$ is the real network Γ. The solution $\Gamma'=(V,c')$ with $\delta=\frac{c(x,y)c(y,z)}{c(x,y)+c(y,z)}$ satisfies $c'(x,y)=c'(y,z)=0$, so it has only one edge, $E(\Gamma')=\{e_{xz}\}$. Moreover, its conductance satisfies

$$\frac{1}{c'(x,z)}=\frac{1}{c(x,y)}+\frac{1}{c(y,z)}, \tag{4}$$

so Γ' is the union of the isolated vertex $\{y\}$ and the Kron reduction of Γ with respect to $\{y\}$; and $c'(x,z)$ is equal to the effective conductance in Γ between x and z, $c^e(x,z)$. The fact that this network is a solution of the problem is because its Laplacian is the Dirichlet-to-Neumann map of Γ and $\{y\}$, which gives us the relationship between potential and injected current at $V\setminus\{y\}=\{x,z\}$ when the potential is harmonic on y. Additionally, the rest of solutions with conductance of form $c'=c+\delta\alpha$, are the infinite solutions with $\delta\in\left(0,\frac{c(x,y)c(y,z)}{c(x,y)+c(y,z)}\right)$, which have the topology of $(V,E(K_V))$, the complete graph in V. Note that Problem 1 can be ill-posed even in cases in which the set of edges E considered is the set of edges of a solution, as happens in this example.

In a numerical context, we usually have errors in the data pair $(\boldsymbol{u},\boldsymbol{s})$, and as a consequence often there is no exact solution to Problem 1. We reformulate the problem in Problem 2, in order to have a solution for any data pair. We seek a network for which the power flow equations are the closest possible to be satisfied in the least squares sense. If $\Gamma=(V,c)$ is a DC network, we define the *error of the network* Γ *in the data pair* $(\boldsymbol{u}, \boldsymbol{s})$, $\mathsf{rms}(\Gamma, \boldsymbol{u}, \boldsymbol{s})$, as the root mean square of the set of evaluations at all vertices of the residuals $u_j\mathcal{L}(u_j) - s_j$ of all the power flow equations corresponding with all the pairs (u_j, s_j), $j = 1, \dots, m$. That is,

$$\mathsf{rms}(\Gamma, \boldsymbol{u}, \boldsymbol{s}) = \frac{||\mathscr{P}_c(\boldsymbol{u}, \boldsymbol{u}) - \boldsymbol{s}||}{\sqrt{m|V|}} = \sqrt{\frac{1}{m|V|}\sum_{j=1}^{m}\int_V \big(u_j\mathcal{L}(u_j) - s_j\big)^2\,dx}.$$

If $\Gamma = (V, a)$ is an AC network, we define the *error of the network* Γ *in the data pair* $(\boldsymbol{u}, \boldsymbol{s})$, $\mathsf{rms}(\Gamma, \boldsymbol{u}, \boldsymbol{s})$, as the root mean square of the set of evaluations at all vertices of the real and imaginary parts of the residuals $u_j\overline{\mathcal{L}(u_j)} - s_j$ of all the power flow equations corresponding with all the pairs (u_j, s_j), $j = 1, \dots, m$. That is,

$$\mathsf{rms}(\Gamma, \boldsymbol{u}, \boldsymbol{s}) = \frac{||\mathscr{P}_a(\boldsymbol{u}, \boldsymbol{u}) - \boldsymbol{s}||}{\sqrt{2m|V|}} = \sqrt{\frac{1}{2m|V|}\sum_{j=1}^{m}\int_V \left|u_j\overline{\mathcal{L}(u_j)} - s_j\right|^2 dx}.$$

Problem 2 *Let V be a finite set of vertices, let E be a set of edges on V, let $0 < |u|_{\min} \le |u|_{\max}$ be positive values, let $m \in \mathbb{N}^*$ and let $\boldsymbol{u}, \boldsymbol{s} \in \mathscr{C}(V, \mathbb{C}^m)$, respectively $\boldsymbol{u}, \boldsymbol{s} \in \mathscr{C}(V, \mathbb{R}^m)$, such that $|u|_{\min} \le |u_j| \le |u|_{\max}$ for all $j = 1, \dots, m$. Determine an AC, respectively a DC, electrical network $\Gamma = (V, a)$ with set of edges $E(\Gamma) \subseteq E$ and such that the error $\mathsf{rms}(\Gamma, \boldsymbol{u}, \boldsymbol{s})$ is minimum.*

Finding a solution to Problem 2 is equivalent to solving a Non-Negative Least Squares (NNLS) problem, that is, a least squares problem with the restriction that the coefficients must be nonnegative, (see [13]). In order to see that, given $\boldsymbol{u} \in \mathscr{C}(V, \mathbb{R}^m)$ and a set of edges E, we define the linear operator $\mathcal{M}_{E,\boldsymbol{u}} : \mathscr{C}(E) \longrightarrow \mathscr{C}(V, \mathbb{R}^m)$ as

$$\mathcal{M}_{E,\boldsymbol{u}}(c) = \big(u_1\mathcal{L}_c(u_1), \dots, u_m\mathcal{L}_c(u_m)\big) = \mathscr{P}_c(\boldsymbol{u}, \boldsymbol{u}).$$

Then, for any DC network $\Gamma = (V, c)$ with set of edges $E(\Gamma) \subseteq E$, and any data pair $(\boldsymbol{u}, \boldsymbol{s}) \in \mathscr{C}(V, \mathbb{R}^m) \times \mathbb{C}(V, \mathbb{R}^m)$, we have that

$$\mathsf{rms}(\Gamma, \boldsymbol{u}, \boldsymbol{s}) = \frac{||\mathcal{M}_{E,\boldsymbol{u}}(c) - \boldsymbol{s}||}{\sqrt{m|V|}}. \tag{5}$$

Similarly, given $\boldsymbol{u} \in \mathscr{C}(V, \mathbb{C}^m)$ and a set of edges E, we define the linear operator $\mathcal{M}_{E,\boldsymbol{u}} : \mathscr{C}(E) \times \mathscr{C}(E) \longrightarrow \mathscr{C}(V, \mathbb{R}^{2m})$ as

$$\mathcal{M}_{E,\boldsymbol{u}}(c, b) = \Big(\Re\big(u_1\overline{\mathcal{L}_a(u_1)}\big), \Im\big(u_1\overline{\mathcal{L}_a(u_1)}\big), \dots, \Re\big(u_m\overline{\mathcal{L}_a(u_m)}\big), \Im\big(u_m\overline{\mathcal{L}_a(u_m)}\big)\Big),$$

where $a = c - ib$.

We also consider the operator $\mathcal{S} : \mathscr{C}(V, \mathbb{C}^m) \longrightarrow \mathscr{C}(V, \mathbb{R}^{2m})$ defined as

$$\mathcal{S}(s) = \Big(\Re(s_1), \Im(s_1), \ldots, \Re(s_m), \Im(s_m)\Big).$$

Then, for any AC network $\Gamma = (V, a)$ with set of edges $E(\Gamma) \subseteq E$, where $a = c - ib$, and any data pair $(\boldsymbol{u}, \boldsymbol{s}) \in \mathscr{C}(V, \mathbb{C}^m) \times \mathbb{C}(V, \mathbb{C}^m)$, we have that

$$\mathsf{rms}(\Gamma, \boldsymbol{u}, \boldsymbol{s}) = \frac{||\mathcal{M}_{E,\boldsymbol{u}}(c, b) - \mathcal{S}(\boldsymbol{s})||}{\sqrt{2m|V|}}.$$

Both for DC and AC networks, Problem 2 consists in seeking for a nonnegative function $c \in \mathscr{C}^+(E)$ in the case of DC networks, or two nonnegative functions $c, b \in \mathscr{C}^+(E)$ in the case of AC networks, such that the norm of the difference between its image under the real linear operator $\mathcal{M}_{E,\boldsymbol{u}}$ and the vector function $\boldsymbol{s}$, or $\mathcal{S}(\boldsymbol{s})$, is minimum, so Problem 2 is equivalent to a NNLS problem.

Any NNLS problem is a convex quadratic optimization problem, (see [13]), so every local minimum of it is a global minimum. We can obtain a solution Γ to Problem 2 calculating a minimum of its associated NNLS problem with an interior point method. The interior point methods are among the main numerical algorithms used to solve NNLS problems, (see [5]). We denote the numerical solution Γ with error $\mathsf{rms} \equiv \mathsf{rms}(\Gamma, \boldsymbol{u}, \boldsymbol{s})$ obtained from E and $(\boldsymbol{u}, \boldsymbol{s})$ as $[\Gamma, \mathsf{rms}] = \mathsf{network\_recovery}(E, \boldsymbol{u}, \boldsymbol{s})$.

A sufficient condition that implies that Problem 2 has a unique solution is that $m \geq \frac{|E|}{|V|}$ and $\kappa(\mathcal{M}_{E,\boldsymbol{u}}) < \infty$, i.e., the *condition number of* $\mathcal{M}_{E,\boldsymbol{u}}$ is finite, (see [13]). We will require in Sect. 7 that the data sets used to test the sparse network recovery algorithm satisfy $m \geq \frac{|E|}{|V|}$. Nevertheless, often the condition $\kappa(\mathcal{M}_{E,\boldsymbol{u}}) < \infty$ is not satisfied, and thus there are multiple solutions to Problem 2; as seen in Example 1, in which there are multiple solutions with zero error.

As we will see in Sect. 7, in cases in which the pair $(\boldsymbol{u}, \boldsymbol{s})$ contains voltage and power data corresponding to an electrical network with some error, the value of the condition number $\kappa(\mathcal{M}_{E,\boldsymbol{u}})$ is finite but it is very high, so Problem 2 has a unique solution but it is severely ill-posed.

3 Reformulation of the Problem: Recovery of a Sparse Electrical Network

As we have discussed in the previous section, given a data data pair $(\boldsymbol{u}, \boldsymbol{s})$ and a set of edges E, the Problem 2 of recovering a network Γ with minimum error such that $E(\Gamma) \subseteq E$ is ill-posed. Because of that, even if there is a solution Γ^* to the exact Problem 1 whose topology is much more sparse than (V, E), i.e., $|E(\Gamma^*)| \ll |E|$, solving Problem 2 we usually get a solution whose topology is (V, E). If the number of edges in E is high, a solution with that topology is not efficient for applications. We are interested in recovering a sparse network such that the fitting error to the data is below a fixed tolerance.

We formulate the following problem of recovering a sparse network. Given a fixed tolerance, we seek to recover a network such that its rms is below this tolerance and none of the networks with the same set of vertices as our network and a subset of its edges has a rms below the tolerance.

Problem 3 *Let V be a finite set of vertices, let E be a set of edges on V, let $0 < |u|_{\min} \leq |u|_{\max}$ be positive values, let $m \in \mathbb{N}^*$, let $\boldsymbol{u}, \boldsymbol{s} \in \mathscr{C}(V, \mathbb{C}^m)$, respectively $\boldsymbol{u}, \boldsymbol{s} \in \mathscr{C}(V, \mathbb{R}^m)$, such that $|u|_{\min} \leq |u_j| \leq |u|_{\max}$ for all $j = 1, \ldots, m$ and let $\mathsf{tol} > 0$ be a tolerance. Determine an AC network $\Gamma = (V, a)$, respectively a DC network $\Gamma = (V, c)$, with set of edges $E(\Gamma) \subseteq E$ such that $\mathsf{rms}(\Gamma, \boldsymbol{u}, \boldsymbol{s}) \leq \mathsf{tol}$ and Γ is "minimal" in the following sense: Given any electrical network $\Gamma' = (V, a')$, respectively $\Gamma' = (V, c')$, with edge set $E(\Gamma')$, we have that*

1. *If $E(\Gamma') = E(\Gamma)$, then $\mathsf{rms}(\Gamma', \boldsymbol{u}, \boldsymbol{s}) \geq \mathsf{rms}(\Gamma, \boldsymbol{u}, \boldsymbol{s})$.*
2. *If $E(\Gamma') \subsetneq E(\Gamma)$, then $\mathsf{rms}(\Gamma', \boldsymbol{u}, \boldsymbol{s}) > \mathsf{tol}$.*

Remark 2 In particular, we have $\mathsf{rms}(\Gamma', \boldsymbol{u}, \boldsymbol{s}) > \mathsf{tol} \geq \mathsf{rms}(\Gamma, \boldsymbol{u}, \boldsymbol{s})$ if $E(\Gamma') \subsetneq E(\Gamma)$. However, notice that there could be an electrical network $E(\Gamma') = (V, a')$ such that $\mathsf{rms}(\Gamma', \boldsymbol{u}, \boldsymbol{s}) < \mathsf{rms}(\Gamma, \boldsymbol{u}, \boldsymbol{s})$ and $E(\Gamma') \not\subset E(\Gamma)$.

A first naive idea to solve Problem 3 could be to recover a network Γ solving Problem 2 with the set of edges E, i.e., $[\Gamma, \mathsf{rms}] = \mathsf{network\_recovery}(E, \boldsymbol{u}, \boldsymbol{s})$, and then to remove from $E(\Gamma)$ the edges whose admittance values are close to zero. However, this approach presents some problematic issues.

First, let Γ^* be a solution to Problem 3, with set of edges $E(\Gamma^*)$. As we will see in Sect. 7, if $E(\Gamma^*)$ is much smaller than E, then usually the condition number of the operator associated with the network recovery Problem 2 using set E, $\kappa(\mathcal{M}_{E,\boldsymbol{u}})$, is much bigger than the condition number of the operator associated with the network recovery Problem 2 using set $E(\Gamma^*)$, $\kappa(\mathcal{M}_{E(\Gamma^*),\boldsymbol{u}})$. Then, the values of the admittance of Γ at the edges of $E \setminus E(\Gamma^*)$ are usually far from zero. For instance, in Example 2, solving the Problem 2 with the edge set $E(K_V) = \{e_{xy}, e_{xz}, e_{yz}\}$, we can get solutions with zero error such that the values of the admittance at all edges of $E(K_V)$ are far from zero, despite the fact that there are solutions with zero error and less than three edges.

Second, in Γ could exist edges with admittance close to zero whose removal would lead to a network which would not correctly fit the data (see Example 3). In order to avoid those problems, the algorithm that we propose in Sect. 6 algorithm uses techniques of spectral sparsification of networks in order to remove edges from networks.

4 Spectral Network Sparsification

Spielman and Teng introduced the notion of spectral sparsification of a real weighted graph, (that is, a DC network), in [16], (see also the subsequent papers [2, 14, 15] on this topic).

Definition 2 ([14]) For $\varepsilon > 0$, we say that a DC network $\Gamma' = (V, c')$ with energy $\mathcal{E}'$ is an ε-approximation of a DC network $\Gamma = (V, c)$ with energy $\mathcal{E}$ if for all $u \in \mathcal{C}(V)$:

$$\frac{1}{1+\varepsilon}\mathcal{E}(u, u) \leq \mathcal{E}'(u, u) \leq (1+\varepsilon)\mathcal{E}(u, u). \tag{6}$$

In [12] the definition of the energy was extended to AC networks. Given an AC network $\Gamma = (V, a)$, its energy is a complex bilinear form, so we can not apply Definition 2 to it. Nevertheless, we can apply this definition separately to its conductance and susceptance networks, $\Gamma_c = (V, c)$ and $\Gamma_b = (V, b)$. We introduce the following definition.

Definition 3 For $\varepsilon > 0$, we say that an AC network Γ' is an ε-approximation of an AC network Γ if the conductance and susceptance networks of Γ' are ε-approximations of the conductance and susceptance networks of Γ, respectively.

In [2] there is a procedure (Algorithm 1) that can be used to construct a sparse approximation $\Gamma' = (V, c')$ of a DC network $\Gamma = (V, c)$ such that $E(\Gamma') \subseteq E(\Gamma)$.

Algorithm 1 $\Gamma' = \mathsf{Sparsify}(\Gamma, \varepsilon)$.

Set $t = 8|V| \cdot log(|V|)/\varepsilon^2$, $c' = 0$, and $E(\Gamma') = \emptyset$.
for each edge $e_{xy} \in E(\Gamma)$ **do**
  Assign to edge e_{xy} a probability p_{xy} proportional to $c(x, y)r^e(x, y)$.
end for
Take t samples independently with replacement from $E(\Gamma)$, and each time the edge e_{xy} is sampled, increase the value of $c'(x, y)$ and $c'(y, x)$ by $c(x, y)/tp_{xy}$, (and as a consequence, add the edge e_{xy} to $E(\Gamma')$ if it is the first time that e_{xy} is sampled).

Algorithm 1 is not guaranteed to produce an ε-approximation of Γ, but in [2] there is a proof of the following theorem:

Theorem 1 (Batson, Spielman, Srivastava, Teng) *Let Γ be a DC network, let $\varepsilon \in \mathbb{R}$, $0 < \varepsilon \leq 1$ and let $\Gamma' = \mathsf{Sparsify}(\Gamma, \varepsilon)$. Then Γ' is an ε-approximation of Γ with probability at least* $1/2$.

Note that the edges that are removed in Algorithm 1 are the ones that are never sampled. The choice of the sampling probabilities in the algorithm has an interesting physical interpretation. By (2), for each edge $e_{xy} \in E$, the probability of sampling it, p_{xy}, is proportional to the dimensionless ratio

$$c(x, y)r^e(x, y) = \frac{c(x, y)}{c^e(x, y)} = \frac{c(x, y)}{c(x, y) + c^e_{\setminus e_{xy}}(x, y)} \in (0, 1],$$

where $c^e_{\setminus e_{xy}}(x, y) \geq 0$ is the effective conductance between x and y in the network $\Gamma \setminus e_{xy}$ obtained from Γ by removing the edge e_{xy}, which is equal to the contribution

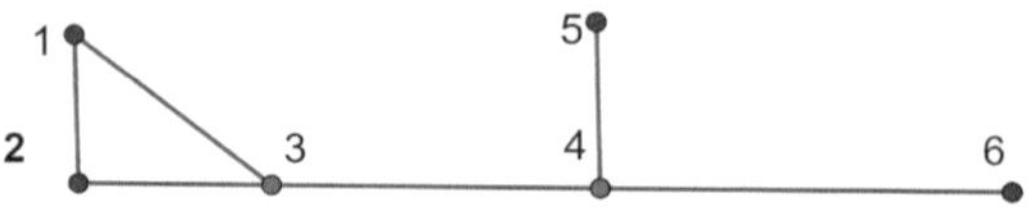

Fig. 2 Topology of the network in Example 2

to the effective conductance $c^e(x, y)$ of the paths between x and y through the rest of vertices of the network Γ.

The probability of choosing edge e_{xy} in each sample is high if the contribution of the value $c(x, y)$ to the effective conductance between x and y in Γ, $c^e(x, y)$, is very important, that is, if $c(x, y)$ is big compared to $c^e_{\setminus e_{xy}}(x, y)$. The limit case is that in which the removal of e_{xy} isolates x and y, in which $c^e_{\setminus e_{xy}}(x, y) = 0$, so $c(x, y)r^e(x, y) = 1$; and thus the probability of keeping e_{xy} in Γ' is the highest possible.

The probability p_{xy} is low in the opposite case, that is, when $c(x, y) \ll c^e_{\setminus e_{xy}}(x, y)$. In that case removal of edge e_{xy} does not significantly affect the effective conductance between x and y, because $c^e_{\setminus e_{xy}}(x, y) \approx c^e(x, y)$. In that case, the probability of keeping e_{xy} in Γ' is low.

Example 2 We consider the DC network in Fig. 2, with conductance c whose values at each edge are given in Table 1. In the same table we indicate the sampling probabilities calculated for each edge in Algorithm 1.

The values of the conductance at edges $\{x_1, x_2\}$ and $\{x_3, x_4\}$ are two orders of magnitude smaller than at the rest of edges, nevertheless, edge $\{x_3, x_4\}$ is crucial because it is the only connection between vertices x_3 and x_4, so it has a great probability of being sampled to form any sparse approximation of the network, while edge $\{x_1, x_2\}$ is expendable, because its conductance $c(x_1, x_2)$ is equal to only 1.5% of the effective conductance between nodes x_1 and x_2. The current can flow from node x_1 to node x_2 with much greater ease by edges $\{x_1, x_3\}$ and $\{x_2, x_3\}$, so the sampling probability of $\{x_1, x_2\}$ is close to zero.

In the case of an AC network $\Gamma = (V, a)$, we will denote by $\Gamma' = \mathsf{Sparsify}(\Gamma, \varepsilon)$ an AC network constructed following the next procedure. First, we apply Algorithm 1 separately to the conductance and susceptance networks, Γ_c and Γ_b, of Γ; getting $\Gamma'_c = (V, c')$ and $\Gamma'_b = (V, b')$, respectively. Then, $\Gamma' = (V, a') = (V, c' - ib')$.

Table 1 Conductances and sampling probabilities of the edges in the network

	Edges					
	$\{x_1, x_2\}$	$\{x_1, x_3\}$	$\{x_2, x_3\}$	$\{x_3, x_4\}$	$\{x_4, x_5\}$	$\{x_4, x_6\}$
Conductance	0.5797	75.980	75.980	0.4698	94.599	79.909
Conductance times effective resistance	0.0150	0.9925	0.9925	1	1	1
Sampling probability	0.0030	0.1985	0.1985	0.2	0.2	0.2

Note that $E(\Gamma') = E(\Gamma'_c) \cup E(\Gamma'_b)$, so in the sparsification process of an AC network we remove the edges erased in the sparsification of both Γ_c and Γ_b.

By definition, Γ' will be an ε-approximation of Γ iff Γ'_c is an ε-approximation of Γ_c and Γ'_b is an ε-approximation of Γ_b. By Theorem 1, if $0 < \varepsilon \leq 1$, then Γ' is an ε-approximation of Γ with probability at least $\frac{1}{4}$.

5 Sparsification of Recovered Electrical Networks

Once we have a procedure to sparsify any network, the key observation to develop our algorithm of sparse network recovery is that spectral sparsification is guaranteed to preserve the fitting properties of a network to some extent.

We introduce some notation for the main result of this work in the case of DC networks. Let $\Gamma = (V, c)$ be a DC network with Laplacian $\mathcal{L}$ and let $\boldsymbol{u}, \boldsymbol{s} \in \mathscr{C}(V, \mathbb{R}^m)$, such that $0 < |u|_{\min} \leq |u_j| \leq |u|_{\max}$ for all $j = 1, \ldots, m$. For each $j = 1, ..., m$, we denote as $\mathscr{Q}_{\Gamma,u_j}$ the linear operator on $\mathscr{C}(V)$ defined for each $v \in \mathscr{C}(V)$ as $\mathscr{Q}_{\Gamma,u_j}(v) = u_j\mathcal{L}(u_j v)$. Clearly, $\mathscr{Q}_{\Gamma,u_j}$ is a self-adjoint and positive semidefinite operator. Now, we denote as $\mathscr{Q}_{\Gamma,\boldsymbol{u}}$ the linear operator on $\mathscr{C}(V, \mathbb{R}^m)$ defined for each $(v_1, ..., v_m) \in \mathscr{C}(V, \mathbb{R}^m)$ as $\mathscr{Q}_{\Gamma,\boldsymbol{u}}(v_1, ..., v_m) = \big(\mathscr{Q}_{\Gamma,u_1}(v_1), ..., \mathscr{Q}_{\Gamma,u_m}(v_m)\big)$. For any $\boldsymbol{v} = (v_1, ..., v_m)$, $\boldsymbol{w} = (w_1, ..., w_m) \in \mathscr{C}(V, \mathbb{R}^m)$, we have that

$$\begin{aligned} \langle \mathscr{Q}_{\Gamma,\boldsymbol{u}}(\boldsymbol{v}), \boldsymbol{w}\rangle &= \textstyle\sum_{j=1}^m \int_V \mathcal{L}(u_j v_j) u_j w_j dx = \sum_{j=1}^m \int_V u_j v_j \mathcal{L}(u_j w_j) dx \\ &= \langle \boldsymbol{v}, \mathscr{Q}_{\Gamma,\boldsymbol{u}}(\boldsymbol{w})\rangle, \end{aligned}$$

so the operator $\mathscr{Q}_{\Gamma,\boldsymbol{u}}$ is self-adjoint. From the same expression, when $\boldsymbol{v} = \boldsymbol{w}$, we can see that $\mathscr{Q}_{\Gamma,\boldsymbol{u}}$ is also positive semidefinite. We denote by $\mathbf{1}_m = (\chi_V, ..., \chi_V) \in \mathscr{C}(V, \mathbb{R}^m)$ the vector function whose components are equal to 1 at all vertices. Note that $\mathscr{Q}_{\Gamma,\boldsymbol{u}}(\mathbf{1}_m) = \mathcal{M}_{E(\Gamma),\boldsymbol{u}}(c)$.

For each $j = 1, ..., m$, we denote as $u_j^{-1} \in \mathscr{C}(V)$ the function defined for each $x \in V$ as $u_j^{-1}(x) = 1/u_j(x)$. We also define the vector function

$$\phi_{\boldsymbol{u}} = \left(\frac{\int_V u_1^{-1} dx}{||u_1^{-1}||^2} u_1^{-1}, \ldots, \frac{\int_V u_m^{-1} dx}{||u_m^{-1}||^2} u_m^{-1} \right) \in \mathscr{C}(V, \mathbb{R}^m).$$

Our main original result consists in the following upper bound for the rms of any sparse approximation in a fixed data set.

Theorem 2 (Main theorem) *Given a DC network $\Gamma = (V, c)$, positive values $0 < |u|_{\min} \leq |u|_{\max}$, $m \in \mathbb{N}^*$ and $\boldsymbol{u}, \boldsymbol{s} \in \mathscr{C}(V, \mathbb{R}^m)$ such that $|u|_{\min} \leq |u_j| \leq |u|_{\max}$ for all $j = 1, \ldots, m$; if Γ' is an ε-approximation of Γ, then:*

$$\mathsf{rms}(\Gamma', \boldsymbol{u}, \boldsymbol{s}) \leq \mathsf{rms}(\Gamma, \boldsymbol{u}, \boldsymbol{s}) + \varepsilon \frac{\|\mathscr{Q}_{\Gamma,\boldsymbol{u}}\|_2 \cdot \|\boldsymbol{1}_m - \phi_{\boldsymbol{u}}\|}{\sqrt{m|V|}}.$$

Proof Let $\Gamma' = (V, c')$ be an ε-approximation of Γ, with Laplacian $\mathcal{L}'$. As $\mathscr{Q}_{\Gamma',\boldsymbol{u}}(\mathbf{1}_m) = \mathcal{M}_{E(\Gamma'),\boldsymbol{u}}(c')$, by (5) we get that

$$\begin{aligned}\mathsf{rms}(\Gamma', \boldsymbol{u}, \boldsymbol{s}) &= \frac{1}{\sqrt{m|V|}}\|\mathscr{Q}_{\Gamma',\boldsymbol{u}}(\mathbf{1}_m) - \boldsymbol{s}\| \\ &\leq \frac{1}{\sqrt{m|V|}}\|\mathscr{Q}_{\Gamma,\boldsymbol{u}}(\mathbf{1}_m) - \boldsymbol{s}\| + \frac{1}{\sqrt{m|V|}}\|\mathscr{Q}_{\Gamma',\boldsymbol{u}}(\mathbf{1}_m) - \mathscr{Q}_{\Gamma,\boldsymbol{u}}(\mathbf{1}_m)\| \\ &= \mathsf{rms}(\Gamma, \boldsymbol{u}, \boldsymbol{s}) + \frac{1}{\sqrt{m|V|}}\|\left(\mathscr{Q}_{\Gamma',\boldsymbol{u}} - \mathscr{Q}_{\Gamma,\boldsymbol{u}}\right)(\mathbf{1}_m)\|.\end{aligned}$$

The constant functions belong to the null space of any Laplacian, therefore $\phi_{\boldsymbol{u}}$ belongs to the null space of $\mathscr{Q}_{\Gamma',\boldsymbol{u}} - \mathscr{Q}_{\Gamma,\boldsymbol{u}}$, and thus

$$\begin{aligned}\|\left(\mathscr{Q}_{\Gamma',\boldsymbol{u}} - \mathscr{Q}_{\Gamma,\boldsymbol{u}}\right)(\mathbf{1}_m)\| &= \|\left(\mathscr{Q}_{\Gamma',\boldsymbol{u}} - \mathscr{Q}_{\Gamma,\boldsymbol{u}}\right)(\mathbf{1}_m - \phi_{\boldsymbol{u}})\| \\ &\leq \|\mathscr{Q}_{\Gamma',\boldsymbol{u}} - \mathscr{Q}_{\Gamma,\boldsymbol{u}}\|_2 \cdot \|\mathbf{1}_m - \phi_{\boldsymbol{u}}\|.\end{aligned} \tag{7}$$

In order to complete the proof, it is enough to show that $\|\mathscr{Q}_{\Gamma',\boldsymbol{u}} - \mathscr{Q}_{\Gamma,\boldsymbol{u}}\|_2 \leq \varepsilon\|\mathscr{Q}_{\Gamma,\boldsymbol{u}}\|_2$. Now, $\mathscr{Q}_{\Gamma',\boldsymbol{u}} - \mathscr{Q}_{\Gamma,\boldsymbol{u}}$ is a self-adjoint operator, so its spectral norm is equal to the maximum of the absolute value of its eigenvalues. It is straightforward to prove that its eigenvalues are those of the operators $\mathscr{Q}_{\Gamma',u_j} - \mathscr{Q}_{\Gamma,u_j}$ for all $1 \leq j \leq m$. In particular, there is at least one index k such that

$$\|\mathscr{Q}_{\Gamma',\boldsymbol{u}} - \mathscr{Q}_{\Gamma,\boldsymbol{u}}\|_2 = \|\mathscr{Q}_{\Gamma',u_k} - \mathscr{Q}_{\Gamma,u_k}\|_2.$$

Let $u_0 \in \mathscr{C}(V)$ be a eigenvector of $\mathscr{Q}_{\Gamma',u_k} - \mathscr{Q}_{\Gamma,u_k}$ corresponding with its eigenvalue that has the largest absolute value such that $\|u_0\| = 1$. Then

$$\|\mathscr{Q}_{\Gamma',\boldsymbol{u}} - \mathscr{Q}_{\Gamma,\boldsymbol{u}}\|_2 = |\langle u_0, \left(\mathscr{Q}_{\Gamma',u_k} - \mathscr{Q}_{\Gamma,u_k}\right)(u_0)\rangle|.$$

Now, denoting as $\mathcal{E}$ and $\mathcal{E}'$ the energy of Γ and Γ', respectively, we have that

$$\begin{aligned}\langle u_0, \left(\mathscr{Q}_{\Gamma',u_k} - \mathscr{Q}_{\Gamma,u_k}\right)(u_0)\rangle &= \langle u_0, u_k(\mathcal{L}' - \mathcal{L})(u_k u_0)\rangle \\ &= \int_V u_0 u_k(\mathcal{L}' - \mathcal{L})(u_0 u_k)dx \\ &= \mathcal{E}'(u_0 u_k, u_0 u_k) - \mathcal{E}(u_0 u_k, u_0 u_k).\end{aligned}$$

Evaluating (6) at $u = u_0 u_k \in \mathscr{C}(V)$ we get

$$\frac{1}{1+\varepsilon}\mathcal{E}(u_0 u_k, u_0 u_k) \leq \mathcal{E}'(u_0 u_k, u_0 u_k) \leq (1+\varepsilon)\mathcal{E}(u_0 u_k, u_0 u_k). \tag{8}$$

From the right inequality of (8), we have:

$$\mathcal{E}'(u_0 u_k, u_0 u_k) - \mathcal{E}(u_0 u_k, u_0 u_k) \leq \varepsilon\mathcal{E}(u_0 u_k, u_0 u_k),$$

and from the left inequality in (8):

$$-\left(\mathcal{E}'(u_0u_k, u_0u_k) - \mathcal{E}(u_0u_k, u_0u_k)\right) \leq \frac{\varepsilon}{1+\varepsilon}\mathcal{E}(u_0u_k, u_0u_k) < \varepsilon\mathcal{E}(u_0u_k, u_0u_k).$$

Joining the last two inequalities, we get

$$\|\mathcal{Q}_{\Gamma',\boldsymbol{u}} - \mathcal{Q}_{\Gamma,\boldsymbol{u}}\|_2 \leq \varepsilon\mathcal{E}(u_0u_k, u_0u_k).$$

Moreover,

$$\begin{aligned}\mathcal{E}(u_0u_k, u_0u_k) &= \int_V u_0u_k\mathcal{L}(u_0u_k)dx\\ &= \langle u_0, u_k\mathcal{L}(u_ku_0)\rangle\\ &= \langle u_0, \mathcal{Q}_{\Gamma,u_k}(u_0)\rangle.\end{aligned}$$

By the Courant-Fisher theorem [15], the evaluation of the quadratic form associated with the operator $\mathcal{Q}_{\Gamma,u_k}$ at any unit vector is less or equal than its maximum eigenvalue, which is equal to the norm of this operator because it is positive semidefinite, so:

$$\|\mathcal{Q}_{\Gamma',\boldsymbol{u}} - \mathcal{Q}_{\Gamma,\boldsymbol{u}}\|_2 \leq \varepsilon\|\mathcal{Q}_{\Gamma,u_k}\|_2 \leq \varepsilon\|\mathcal{Q}_{\Gamma,\boldsymbol{u}}\|_2, \tag{9}$$

where the last inequality holds because the eigenvalues of $\mathcal{Q}_{\Gamma,\boldsymbol{u}}$ are those of the operators $\mathcal{Q}_{\Gamma,u_j}$ for all $1 \leq j \leq m$. □

Remark 3 In (7), it is possible to introduce any vector function equal to $\left(\lambda_1u_1^{-1}, ..., \lambda_mu_m^{-1}\right) \in \mathscr{C}(V, \mathbb{R}^m)$, for any choice of $\lambda_1, ..., \lambda_m \in \mathbb{R}$, because all of them belong to the null space of $\mathcal{Q}_{\Gamma',\boldsymbol{u}} - \mathcal{Q}_{\Gamma,\boldsymbol{u}}$. Among them, we choose to introduce the term $\phi_{\boldsymbol{u}}$ because it is the choice that minimizes $\rho(\lambda_1, .., \lambda_m) \equiv \left\|\mathbf{1}_m - \left(\lambda_1u_1^{-1}, ..., \lambda_mu_m^{-1}\right)\right\|^2 = \sum_{j=1}^m \int_V \left(\chi_V - \lambda_ju_j^{-1}\right)^2 dx$. Effectively, if we look for a minimum of ρ, its partial derivatives must be zero:

$$\frac{\partial\rho(\lambda_1, .., \lambda_m)}{\partial\lambda_k} = 2\int_V \left(-u_k^{-1} + \lambda_k\left(u_k^{-1}\right)^2\right)dx = 0,$$

so the only possible choice for the λ_k is:

$$\lambda_k = \frac{\int_V u_k^{-1}dx}{||u_k^{-1}||^2},$$

which is indeed the global minimum of ρ because its Hessian is positive definite: if $j \neq k$, then $\frac{\partial^2\rho(\lambda_1,...,\lambda_m)}{\partial\lambda_k\partial\lambda_j} = 0$, and for each $k = 1, ..., m$,

$$\frac{\partial^2\rho(\lambda_1, .., \lambda_m)}{\partial\lambda_k^2} = 2\int_V \left(u_k^{-1}\right)^2 dx > 0.$$

Remark 4 The term $\frac{\|\mathcal{Q}_{\Gamma,\boldsymbol{u}}\|_2 \cdot \|\mathbf{1}_m - \phi_{\boldsymbol{u}}\|}{\sqrt{m|V|}}$ that appears in the upper bound given by the last theorem depends only on the electrical network Γ and the first element of the data pair $(\boldsymbol{u}, \boldsymbol{s})$, so if we want a sparse approximation of Γ such that the rms of the approximation does not increase (with respect to $\mathsf{rms}\,(\Gamma, \boldsymbol{u}, \boldsymbol{s})$) in the sparsification procedure above a fixed value, there is an $\varepsilon > 0$ such that any ε-approximation of Γ is guaranteed to meet this requirement.

In the case of an AC electrical network Γ, we denote by $\Gamma_c = (V, c)$ and $\Gamma_b = (V, b)$ the conductance and susceptance networks of Γ, whose respective Laplacians are $\mathcal{L}_c$ and $\mathcal{L}_b$. Additionally, we define the function

$$\Delta(\Gamma, \boldsymbol{u}) = \Bigg(\Big(\sqrt{m|V|}(\|\mathcal{Q}_{\Gamma_c,\Re(\boldsymbol{u})}\|_2 + \|\mathcal{Q}_{\Gamma_c,\Im(\boldsymbol{u})}\|_2) + \|\mathcal{L}_b\|_2\,(\|\Re(\boldsymbol{u})\|_\infty \|\Im(\boldsymbol{u})\|_2 + \|\Im(\boldsymbol{u})\|_\infty \|\Re(\boldsymbol{u})\|_2)\Big)^2 + \\ + \Big(\sqrt{m|V|}(\|\mathcal{Q}_{\Gamma_b,\Re(\boldsymbol{u})}\|_2 + \|\mathcal{Q}_{\Gamma_b,\Im(\boldsymbol{u})}\|_2) + \|\mathcal{L}_c\|_2(\|\Im(\boldsymbol{u})\|_\infty \|\Re(\boldsymbol{u})\|_2 + \|\Re(\boldsymbol{u})\|_\infty \|\Im(\boldsymbol{u})\|_2)\Big)^2\Bigg)^{1/2},$$

and we have the following result, which is the AC analogous to Theorem 2:

Theorem 3 *Given an AC network* $\Gamma = (V, a)$, $m \in \mathbb{N}^*$ *and* $\boldsymbol{u}, \boldsymbol{s} \in \mathcal{C}(V, \mathbb{C}^m)$, *if* Γ' *is an* ε*-approximation of* Γ, *then:*

$$\mathsf{rms}(\Gamma', \boldsymbol{u}, \boldsymbol{s}) \leq \mathsf{rms}(\Gamma, \boldsymbol{u}, \boldsymbol{s}) + \varepsilon \frac{\Delta(\Gamma, \boldsymbol{u})}{\sqrt{2m|V|}},$$

where $\Delta(\Gamma, \boldsymbol{u})$ *is a function which depends only on* Γ *and* $\boldsymbol{u}$.

Proof First, for every $u \in \mathcal{C}(V, \mathbb{C})$, we have that

$$u\overline{\mathcal{L}(u)} = u\mathcal{L}^*(\overline{u}) = (\Re(u) + i\Im(u))(\mathcal{L}_c + i\mathcal{L}_b)((\Re(u) - i\Im(u))),$$

and therefore

$$\Re(u\overline{\mathcal{L}(u)}) = \Re(u)\mathcal{L}_c(\Re(u)) + \Re(u)\mathcal{L}_b(\Im(u)) + \Im(u)\mathcal{L}_b(\Re(u)) + \Im(u)\mathcal{L}_c(\Im(u)), \tag{10}$$

and also

$$\Im(u\overline{\mathcal{L}(u)}) = \Im(u)\mathcal{L}_c(\Re(u)) + \Im(u)\mathcal{L}_b(\Im(u)) + \Re(u)\mathcal{L}_b(\Re(u)) + \Re(u)\mathcal{L}_c(\Im(u)). \tag{11}$$

Now, let $\Gamma' = (V, a')$ be an ε-approximation of Γ, with Laplacian $\mathcal{L}'$. Let $\Gamma'_c = (V, c')$ and $\Gamma'_b = (V, b')$ be the conductance and susceptance networks of Γ', whose respective Laplacians are $\mathcal{L}'_c$ and $\mathcal{L}'_b$. By (5) we get that

$$\begin{aligned}\mathsf{rms}(\Gamma', \boldsymbol{u}, \boldsymbol{s}) &= \frac{1}{\sqrt{m|V|}}\|\mathcal{M}_{E(\Gamma'),\boldsymbol{u}}(c', b') - \mathcal{S}(\boldsymbol{s})\| \\ &\le \frac{1}{\sqrt{m|V|}}\|\mathcal{M}_{E(\Gamma),\boldsymbol{u}}(c, b) - \mathcal{S}(\boldsymbol{s})\| \\ &+ \frac{1}{\sqrt{m|V|}}\|\mathcal{M}_{E(\Gamma'),\boldsymbol{u}}(c', b') - \mathcal{M}_{E(\Gamma),\boldsymbol{u}}(c, b)\| \\ &= \mathsf{rms}(\Gamma, \boldsymbol{u}, \boldsymbol{s}) + \frac{1}{\sqrt{m|V|}}\|\mathcal{M}_{E(\Gamma'),\boldsymbol{u}}(c', b') - \mathcal{M}_{E(\Gamma),\boldsymbol{u}}(c, b)\|.\end{aligned}$$

The square of the norm in the last equation is equal to

$$\begin{aligned}&\|\mathcal{M}_{E(\Gamma'),\boldsymbol{u}}(c', b') - \mathcal{M}_{E(\Gamma),\boldsymbol{u}}(c, b)\|^2 \\ &= \|\big(\Re(u_1\overline{(\mathcal{L}' - \mathcal{L})\,(u_1)}), ..., \Re(u_m\overline{(\mathcal{L}' - \mathcal{L})\,(u_m)})\big)\|^2 \\ &+ \|\big(\Im(u_1\overline{(\mathcal{L}' - \mathcal{L})\,(u_1)}), ..., \Im(u_m\overline{(\mathcal{L}' - \mathcal{L})\,(u_m)})\big)\|^2. \end{aligned} \tag{12}$$

By Eqs. (10), (11) and (12), we can write

$$\begin{aligned}\|\mathcal{M}_{E(\Gamma'),\boldsymbol{u}}(c', b') - \mathcal{M}_{E(\Gamma),\boldsymbol{u}}(c, b)\|^2 &= \|\beta_1 + \beta_2 - \beta_3 + \beta_4\|^2 \\ &+ \|\beta_5 + \beta_6 + \beta_7 - \beta_8\|^2 \\ &\le (\|\beta_1\| + \|\beta_2\| + \|\beta_3\| + \|\beta_4\|)^2 \\ &+ (\|\beta_5\| + \|\beta_6\| + \|\beta_7\| + \|\beta_8\|)^2\,,\end{aligned} \tag{13}$$

where:

$$\begin{aligned}\beta_1 &= (\mathscr{P}_{c'} - \mathscr{P}_c)\,(\Re(\boldsymbol{u}), \Re(\boldsymbol{u}))\,, & \beta_2 &= (\mathscr{P}_{b'} - \mathscr{P}_b)\,(\Re(\boldsymbol{u}), \Im(\boldsymbol{u}))\,, \\ \beta_3 &= (\mathscr{P}_{b'} - \mathscr{P}_b)\,(\Im(\boldsymbol{u}), \Re(\boldsymbol{u}))\,, & \beta_4 &= (\mathscr{P}_{c'} - \mathscr{P}_c)\,(\Im(\boldsymbol{u}), \Im(\boldsymbol{u}))\,, \\ \beta_5 &= (\mathscr{P}_{c'} - \mathscr{P}_c)\,(\Im(\boldsymbol{u}), \Re(\boldsymbol{u}))\,, & \beta_6 &= (\mathscr{P}_{b'} - \mathscr{P}_b)\,(\Im(\boldsymbol{u}), \Im(\boldsymbol{u}))\,, \\ \beta_7 &= (\mathscr{P}_{b'} - \mathscr{P}_b)\,(\Re(\boldsymbol{u}), \Re(\boldsymbol{u}))\,, & \beta_8 &= (\mathscr{P}_{c'} - \mathscr{P}_c)\,(\Re(\boldsymbol{u}), \Im(\boldsymbol{u}))\,.\end{aligned}$$

Then, on one hand, $\mathscr{P}_c\,(\Re(\boldsymbol{u}), \Re(\boldsymbol{u})) = \mathscr{Q}_{\Gamma_c,\Re(\boldsymbol{u})}(\mathbf{1}_m)$ and $\mathscr{P}_{c'}\,(\Re(\boldsymbol{u}), \Re(\boldsymbol{u})) = \mathscr{Q}_{\Gamma_c',\Re(\boldsymbol{u})}(\mathbf{1}_m)$, so by (9), we get

$$\begin{aligned}\|\beta_1\| &\le \|\mathscr{Q}_{\Gamma_c',\Re(\boldsymbol{u})} - \mathscr{Q}_{\Gamma_c,\Re(\boldsymbol{u})}\|_2 \cdot \|\mathbf{1}_m\| = \sqrt{m|V|}\|\mathscr{Q}_{\Gamma_c',\Re(\boldsymbol{u})} - \mathscr{Q}_{\Gamma_c,\Re(\boldsymbol{u})}\|_2 \\ &\le \varepsilon\sqrt{m|V|}\|\mathscr{Q}_{\Gamma_c,\Re(\boldsymbol{u})}\|_2.\end{aligned}$$

Reasoning analogously, we also obtain that $\|\beta_4\| \le \varepsilon\sqrt{m|V|}\|\mathscr{Q}_{\Gamma_c,\Im(\boldsymbol{u})}\|_2$, $\|\beta_6\| \le \varepsilon\sqrt{m|V|}\|\mathscr{Q}_{\Gamma_b,\Im(\boldsymbol{u})}\|_2$ and $\|\beta_7\| \le \varepsilon\sqrt{m|V|}\|\mathscr{Q}_{\Gamma_b,\Re(\boldsymbol{u})}\|_2$.

On the other hand, $\mathcal{L}_c = \mathscr{Q}_{\Gamma_c,\chi_V}$, $\mathcal{L}_c' = \mathscr{Q}_{\Gamma_c',\chi_V}$, $\mathcal{L}_b = \mathscr{Q}_{\Gamma_b,\chi_V}$ and $\mathcal{L}_b' = \mathscr{Q}_{\Gamma_b',\chi_V}$, so, by the proof of Theorem 2, we have that $\|\mathcal{L}_c' - \mathcal{L}_c\|_2 \le \varepsilon\|\mathcal{L}_c\|_2$ and $\|\mathcal{L}_b' -$

$\mathcal{L}_b\|_2 \leq \varepsilon \|\mathcal{L}_b\|_2$. As a consequence, for any $(\boldsymbol{v}, \boldsymbol{w}) = ((v_1, ..., v_m), (w_1, ..., w_m)) \in \mathscr{C}(V, \mathbb{R}^m) \times \mathscr{C}(V, \mathbb{R}^m)$ it is satisfied that

$$\begin{aligned}
\|(\mathscr{P}_{c'} - \mathscr{P}_c)(\boldsymbol{v}, \boldsymbol{w})\|^2 &= \sum_{j=1}^{m} \int_V v_j^2 (\mathcal{L}_c' - \mathcal{L}_c)(w_j)^2 dx \\
&\leq \|\boldsymbol{v}\|_\infty^2 \sum_{j=1}^{m} \|(\mathcal{L}_c' - \mathcal{L}_c)(w_j)\|_2^2 \\
&\leq \|\boldsymbol{v}\|_\infty^2 \cdot \|(\mathcal{L}_c' - \mathcal{L}_c)\|_2^2 \cdot \|\boldsymbol{w}\|_2^2 \\
&\leq \varepsilon^2 \|\mathcal{L}_c\|_2^2 \cdot \|\boldsymbol{v}\|_\infty^2 \cdot \|\boldsymbol{w}\|_2^2.
\end{aligned}$$

And, analogously, it is satisfied that $\|(\mathscr{P}_{b'} - \mathscr{P}_b)(\boldsymbol{v}, \boldsymbol{w})\| \leq \varepsilon \|\mathcal{L}_b\|_2 \cdot \|\boldsymbol{v}\|_\infty \cdot \|\boldsymbol{w}\|_2$. Applying this result, we get the following bounds:

$$\|\beta_2\| \leq \|\mathcal{L}_b\|_2 \cdot \|\Re(\boldsymbol{u})\|_\infty \cdot \|\Im(\boldsymbol{u})\|_2, \qquad \|\beta_3\| \leq \|\mathcal{L}_b\|_2 \cdot \|\Im(\boldsymbol{u})\|_\infty \cdot \|\Re(\boldsymbol{u})\|_2,$$
$$\|\beta_5\| \leq \|\mathcal{L}_c\|_2 \cdot \|\Im(\boldsymbol{u})\|_\infty \cdot \|\Re(\boldsymbol{u})\|_2 \quad \text{and} \quad \|\beta_8\| \leq \|\mathcal{L}_c\|_2 \cdot \|\Re(\boldsymbol{u})\|_\infty \cdot \|\Im(\boldsymbol{u})\|_2.$$

Considering in (13) all the upper bounds obtained for $\|\beta_1\|, ..., \|\beta_8\|$, we conclude the proof. □

6 Algorithm for Sparse Network Recovery

In this section we propose an algorithm to solve the Problem 3 of recovering simultaneously the topology and admittance of a sparse electrical network.

If we have an electrical network $\Gamma = (V, a)$, a data pair $(\boldsymbol{u}, \boldsymbol{s})$ and we choose an ε such that the rms of any ε-approximation of Γ in the data pair $(\boldsymbol{u}, \boldsymbol{s})$ does not surpass a fixed tolerance $\mathsf{tol} > 0$ (by Theorems 2 and 3 such a ε exists if $\mathsf{rms}(\Gamma, \boldsymbol{u}, \boldsymbol{s}) < \mathsf{tol}$), the only guarantee about the number of edges that any ε-approximation obtained by executing $\mathsf{Sparsify}(\Gamma, \varepsilon)$ will have is that this number of edges will be less or equal than the number of edges sampled in Algorithm 1, $t = 8|V| \cdot \log(|V|)/\varepsilon^2$. If tol is small, the number of edges sampled will be large, so most ε-approximations of Γ will have the same number of edges as the original network, thus they will not be useful for our purposes.

In the experimentation we have found that, if we choose an ε such that $\Gamma' = \mathsf{Sparsify}(\Gamma, \varepsilon)$ satisfies that $|E(\Gamma')| < |E(\Gamma)|$, and then we solve Problem 2 with the set $E(\Gamma')$; (that is, we execute $[\Gamma'', \mathsf{rms}''] = \mathsf{network\_recovery}(E(\Gamma'), \boldsymbol{u}, \boldsymbol{s})$, determining a network $\Gamma'' = (V, a'')$ with set of edges $E(\Gamma'') \subseteq E(\Gamma')$ such that the error $\mathsf{rms}'' = \mathsf{rms}(\Gamma'', \boldsymbol{u}, \boldsymbol{s})$ is minimum), then we have $\mathsf{rms}'' \leq \mathsf{tol}$ in many cases, even in cases in which $\mathsf{rms}(\Gamma', \boldsymbol{u}, \boldsymbol{s}) > \mathsf{tol}$.

Our approach to solve Problem 3 consists in the application of Algorithm 2. The algorithm starts by fitting a network $\Gamma = \mathsf{network\_recovery}(E, \boldsymbol{u}, \boldsymbol{s})$ solving Problem 2 with the data pair $(\boldsymbol{u}, \boldsymbol{s})$ and the set of edges E. Then, our goal is to perform a sparsification of Γ, $\Gamma' = \mathsf{Sparsify}(\Gamma, \varepsilon)$, followed by a recovery of another network

Γ'' solving Problem 2 with the set of edges of the sparse approximation, $E(\Gamma')$, as in the paragraph above, looking for a network with less edges than the current one, and with a rms below the input tolerance tol.

We do not know which choices of ε can lead us to a network with these characteristics, so we use a procedure of exploration to let Algorithm 2 find suitable values of ε. We start the algorithm with an initial input value of ε, and we do an iterative procedure. Each iteration begins by checking if $\Gamma' = \mathsf{Sparsify}(\Gamma, \varepsilon)$ has less edges than the current network Γ. If this is not the case, there is a high probability, from Theorem 1, of Γ' being a sparse approximation so close to Γ that it has the same set of edges, so we increase the value of ε multiplying it by the input parameter $\psi > 1$, and we finish the iteration. In this way, in the next iteration the number of edges sampled in $\mathsf{Sparsify}$ will be lower than in the previous one, increasing the probability of getting sparse approximations with less edges than Γ.

If, on the contrary, $|E(\Gamma')| < |E(\Gamma)|$, we recover the network Γ'' solving Problem 2 with set of edges $E(\Gamma')$, that is, we compute $[\Gamma'', \mathsf{rms}''] = \mathsf{network\_recovery}(E(\Gamma'), \boldsymbol{u}, \boldsymbol{s})$.

Then, we check if $\mathsf{rms}'' \leq \mathsf{tol}$. If this is the case, we replace the current network Γ by this new Γ'' and we finish the iteration, using the new network as input for $\mathsf{Sparsify}$ in the next iteration, repeating the process, in order to look for networks with less edges than it. If, on the opposite case, $\mathsf{rms}'' > \mathsf{tol}$, this means we have removed edges that are necessary for the network to correctly fit the data in the sparsification process, because there are no networks with set of edges contained in $E(\Gamma')$ whose rms is lower or equal than tol. We reject the network Γ'', we decrease the value of ε dividing it by ψ and we finish the iteration because, from Theorems 2 and 3, we know that decreasing the value of ε assures that the rms on any subsequent ε-approximation of Γ will have a smaller upper bound.

It is possible to define different stopping criteria for Algorithm 2, such as the total running time if we are interested on the best network that the algorithm can find in that period, or a maximum number of consecutive iterations in which the network has not changed. In the case of distribution networks, it is usual that the position of the switches is set so that the network topology is a tree, so in this case, reaching a tree topology can be another stopping criterion. A detailed discussion about the stopping criteria is left for future work.

7 Experimental Results and Discussion

Lastly, we present some results of the application of Algorithm 2 to solve examples of Problem 3. The algorithm has been written in MATLAB using Casadi [19], an open-source software tool that provides a symbolic framework suited for numerical optimization, and the interior-point solver IPOPT [20], an open-source software package for large-scale nonlinear optimization, for the $\mathsf{network\_recovery}$ process. The tolerance used in IPOPT is equal to 10^{-8}. In all the examples, we use $\psi = 1.5$ as input of Algorithm 2.

Algorithm 2 Sparse network recovery.

Inputs: $(\boldsymbol{u}, \boldsymbol{s})$, E, ε, $\psi > 1$, tol, $total\_time$.
$[\Gamma, \text{rms}] := \text{network\_recovery}(E, \boldsymbol{u}, \boldsymbol{s})$.
while $running\_time < total\_time$ **do** // (Stopping criteria)
  Set $\Gamma' := \text{Sparsify}(\Gamma, \varepsilon)$.
  if $|E(\Gamma')| < |E(\Gamma)|$ **then**
    $[\Gamma'', \text{rms}''] = \text{network\_recovery}(E(\Gamma'), \boldsymbol{u}, \boldsymbol{s})$.
    if $\text{rms}'' \leq \text{tol}$ **then**
      $\Gamma := \Gamma''$.
    else
      $\varepsilon := \varepsilon/\psi$.
    end if
  else
    $\varepsilon = \varepsilon \cdot \psi$.
  end if
end while
Return Γ.

Table 2 Example 3 network fitting results with tolerance 10^{-5}

Iteration number	$\|E(\Gamma)\|$	$\text{rms}(\Gamma, \boldsymbol{u}, \boldsymbol{s})$	$\kappa(\mathcal{M}_{E(\Gamma),\boldsymbol{u}})$	ε
1	15	$1.011 \cdot 10^{-6}$	$1.040 \cdot 10^{4}$	0.1000
2	9	$9.415 \cdot 10^{-7}$	$4.316 \cdot 10^{3}$	0.1000
3	7	$4.632 \cdot 10^{-7}$	$3.021 \cdot 10^{3}$	0.1000
4	6	$1.491 \cdot 10^{-8}$	$2.100 \cdot 10^{3}$	0.1000

In each example, we have chosen a network Γ and we have sampled a data pair $(\boldsymbol{u}, \boldsymbol{s})$ consisting in $m = 1000$ pairs of voltage and power injected at Γ. Each pair (u_j, s_j) has been computed numerically solving the power flow equations of Γ, $s_j = u_j\overline{\mathcal{L}(u_j)}$, along with different additional restrictions. Restrictions that are common for all the pairs (u_j, s_j), such as for instance having $s_j(x) = 0$ for a vertex $x \in V$ for all $i = 1, ...m$, will be highlighted only if they are relevant for the analysis of the results.

In all the examples, we solve Problem 3 using $E = E(K_V)$ as set of edges, that is, we solve the problem without any a priori information about the topology of the network. We fix a labeling $\{x_1, ..., x_n\}$ on the set of vertices of each network and in the figures we denote each vertex x_j as j for the sake of clarity.

Example 3 We start with the DC network in Example 2, (whose conductance is shown in Table 1). In the data pair $(\boldsymbol{u}, \boldsymbol{s})$, for each $j = 1, ..., m$ we have different loads, (that is, negative power injected s_j) at all nodes, except at node x_1, where the power feeding the entire network is generated ($s_j(x_1) \geq 0$). We apply Algorithm 2 with $\text{tol} = 10^{-5}$ and an initial value for ε equal to 0.1 and we obtain the results of Table 2 (we only write the results of iterations in which the number of edges has decreased, and we consider the first network recovery with the set of edges $E(K_V)$ of the complete graph previous to the iterative process as the iteration number 1).

Table 3 Example 3 network fitting results with tolerance 10^{-3}

Iteration number	$\lvert E(\Gamma)\rvert$	$\mathsf{rms}(\Gamma, \boldsymbol{u}, \boldsymbol{s})$	$\kappa(\mathcal{M}_{E(\Gamma),\boldsymbol{u}})$	ε
1	15	$1.011 \cdot 10^{-6}$	$1.040 \cdot 10^{4}$	0.1000
2	11	$8.418 \cdot 10^{-7}$	$6.354 \cdot 10^{3}$	0.1000
3	6	$3.580 \cdot 10^{-5}$	$1.320 \cdot 10^{3}$	0.1000
5	5	$3.579 \cdot 10^{-5}$	$7.028 \cdot 10^{2}$	0.1500

The last row in Table 2 corresponds to the final network found by the algorithm, that has the real topology of the network (Fig. 2). The difference between the values at each edge of the conductance obtained and of the real conductance is of the order of magnitude of 10^{-4}. We obtain almost the real network in just 4 iterations.

If we repeat the experiment with a looser tolerance $\mathsf{tol} = 10^{-3}$, we obtain the results in Table 3. The final network is a sparse approximation of the original, with the same edges than it except for edge $\{x_1, x_2\}$, that is removed. This removal barely distorts the structure of the network, as we discussed in Example 2. It is important to remark that the algorithm never removes the low conductance edge $\{x_3, x_4\}$. That removal would make impossible satisfying the demand of power at nodes x_4, x_5 and x_6, because node x_1 is the only node with generated power, so the rms of the resulting network would be higher than tol.

From now on, we will use $\mathsf{tol} = 10^{-5}$ for all experiments.

Example 4 The next experiment uses the CIGRE test AC network named "Medium voltage distribution network with PV and Wind DER" [7] (Fig. 3).

An application of Algorithm 2 gives the results in Table 4.

As we can see in Fig. 4, in this example, the sparsification parameter starts with a value of $\varepsilon = 0.3$, and succeeds at eliminating most edges in the first few iterations. After getting a network with 15 edges, there are some iterations in which Algorithm 1 does not remove any edge. Therefore, Algorithm 2 looks for looser sparse approximations of the network increasing the value of ε, but decreasing its value when a sparse approximation does not fit the data up to the desired tolerance; until at iteration number 15. In this iteration the algorithm arrives to the real topology, with almost the real values of the admittance. After this iteration, the algorithm never removes any more edges, and the value of ε tends to oscillate, taking values between 2.28 and 7.69. In Fig. 4, the iterations in which edges were removed are marked with a circle.

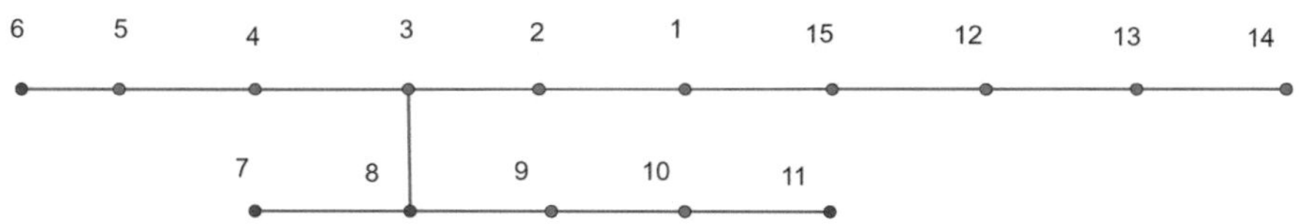

Fig. 3 Topology of the CIGRE network. (Example 4)

Table 4 Example 4 network fitting results

Iteration number	$\lvert E(\Gamma)\rvert$	$\mathsf{rms}(\Gamma, \boldsymbol{u}, \boldsymbol{s})$	$\kappa(\mathcal{M}_{E(\Gamma),\boldsymbol{u}})$	ε
1	105	$2.905 \cdot 10^{-5}$	$4.200 \cdot 10^{15}$	0.3000
2	24	$8.772 \cdot 10^{-6}$	$1.492 \cdot 10^{15}$	0.3000
3	18	$6.405 \cdot 10^{-6}$	$1.526 \cdot 10^{15}$	0.3000
4	17	$6.397 \cdot 10^{-6}$	$1.497 \cdot 10^{15}$	0.3000
6	16	$4.343 \cdot 10^{-6}$	$1.482 \cdot 10^{15}$	0.4500
8	15	$1.668 \cdot 10^{-8}$	$1.476 \cdot 10^{15}$	0.6750
15	14	$1.668 \cdot 10^{-8}$	$7.965 \cdot 10^{2}$	3.417

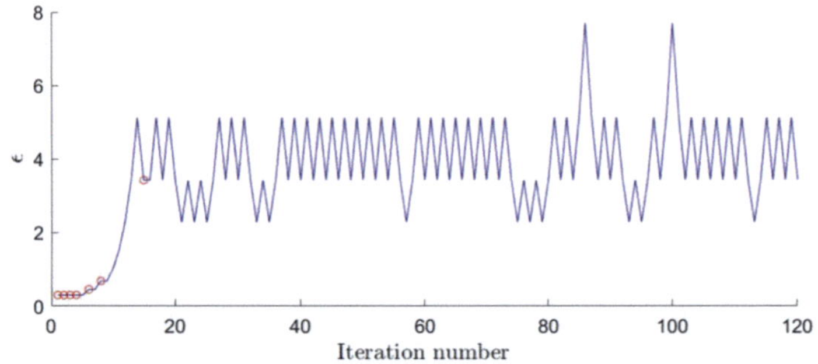

Fig. 4 Evolution of ε with the iteration number. Example 4

Table 5 Heawood graph network fitting results

Iteration number	$\lvert E(\Gamma)\rvert$	$\mathsf{rms}(\Gamma, \boldsymbol{u}, \boldsymbol{s})$	$\kappa(\mathcal{M}_{E(\Gamma),\boldsymbol{u}})$	ε
1	91	$3.705 \cdot 10^{-6}$	$2.119 \cdot 10^{4}$	0.1000
2	35	$9.267 \cdot 10^{-7}$	$2.472 \cdot 10^{3}$	0.1000
3	24	$6.378 \cdot 10^{-7}$	$5.365 \cdot 10^{2}$	0.1000
4	21	$9.164 \cdot 10^{-10}$	$2.586 \cdot 10^{2}$	0.1000

Example 5 In the next experiment, we use a DC network with the topology of the Heawood graph (Fig. 5). The results of the application of Algorithm 2 with an initial value of ε equal to 0.1 are shown in Table 5. The algorithm is able to find the topology of the real network in just 5 iterations in this case.

In Table 6 we compare the rms of the sparse approximation in each iteration that removes any edge with the upper bound on this rms given by Theorem 2.

The rms of the sparse approximations are an order of magnitude smaller than the upper bound guaranteed by Theorem 2. The network recovery step that follows any sparsification step decreases the rms of the network various orders of magnitude. It helps at keeping the rms of the network below tol and lets us avoid the possibility of accumulation of error on successive sparsification steps, which could lead to a network that does not correctly fit the data.

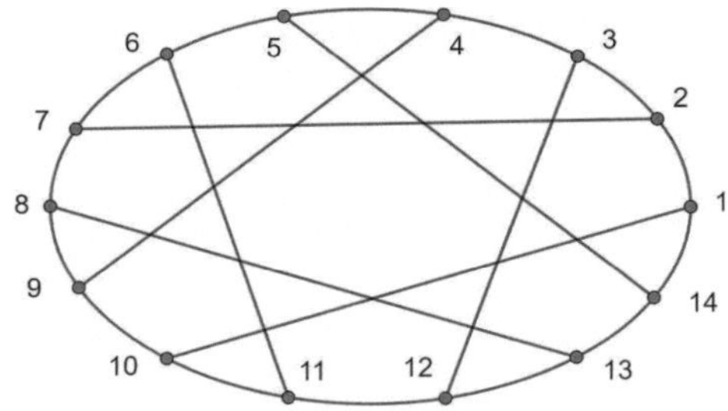

Fig. 5 Iteration number: 5, $|E(\Gamma)| = 21$. (Heawood graph network)

Table 6 Heawood graph rms on the sparse approximations

Iteration number	Upper bound on the rms given by Theorem 2	rms of the sparse approximation
2	0.4429	0.0369
3	0.4432	0.0308
4	0.4432	0.0388
5	0.4432	0.0318

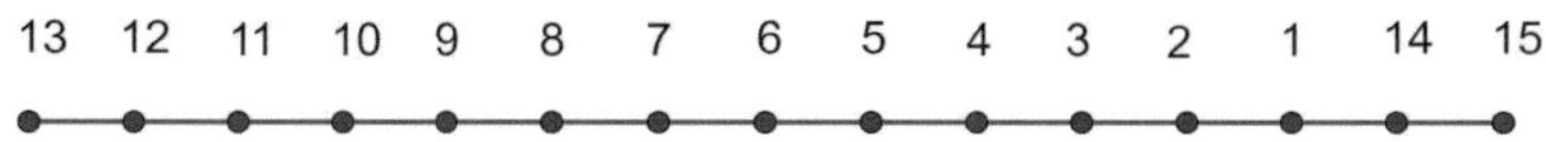

Fig. 6 Landnetz Freileitung 1 network

Table 7 Kerber network fitting results

Iteration number	$\|E(\Gamma)\|$	rms $(\Gamma, \boldsymbol{u}, \boldsymbol{s})$	$\kappa(\mathcal{M}_{E(\Gamma),\boldsymbol{u}})$	ε
1	105	$2.069 \cdot 10^{-7}$	$1.488 \cdot 10^{15}$	0.3
2	16	$2.757 \cdot 10^{-7}$	$3.829 \cdot 10^{14}$	0.3
3	15	$2.485 \cdot 10^{-12}$	$3.825 \cdot 10^{14}$	0.3
10	14	$3.885 \cdot 10^{-7}$	$1.044 \cdot 10^{2}$	3.417
34	13	$2.288 \cdot 10^{-12}$	$1.029 \cdot 10^{2}$	5.126

Example 6 Finally, we present an example of sparse network recovery on the AC low voltage network called “Landnetz Freileitung 1 network” (Fig. 6), that belongs to the Kerber test networks [8].

In the data pair $(\boldsymbol{u}, \boldsymbol{s})$ there is no injected power at node x_{14} for any component of $\boldsymbol{s}$, that is, $s_j(x_{14}) = 0$ and thus the potential u_j is harmonic on x_{14}, for all $j = 1, ..., m$. We apply Algorithm 2 with an initial value of ε equal to 0.3; which gives the results in Table 7.

The final network $\Gamma' = (V, a')$ has one edge less than the real one $\Gamma = (V, a)$, and has the following topology, which is shown in Fig. 7: the edges $\{x_1, x_{14}\}$ and $\{x_{14}, x_{15}\}$ of the real network are substituted in the recovered network by edge $\{x_1, x_{15}\}$. The values of admittance on these edges in the real network Γ are $a(x_1, x_{14}) = 16.7913 -$

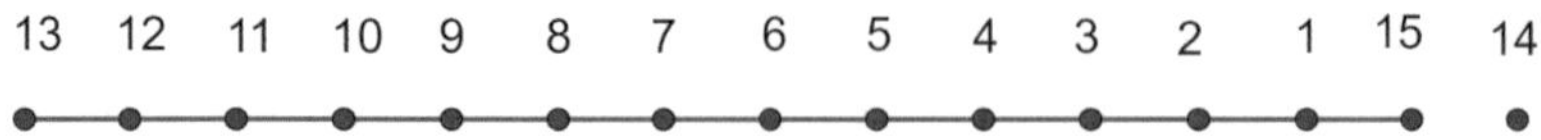

Fig. 7 Iteration number: 34, $|E(\Gamma)| = 13$. (Kerber network)

2.6154i and $a(x_{14}, x_{15}) = 1.1999 - 3.8157i$, and the value of the admittance of Γ' at edge $\{x_1, x_{15}\}$ is $a'(x_1, x_{15}) = 1.6852 - 3.1333i$. The values of a and a' at the edges that are common to both networks, Γ and Γ', are practically equal.

The difference between Γ and Γ' is analogous to the difference in Example 4 between the real network of that example whose set of edges is $\left\{e_{xy}, e_{yz}\right\}$ and the recovered network of that example whose set of edges is $\{e_{xz}\}$. That is, it is satisfied that

$$\frac{1}{a'(x_1, x_{15})} = \frac{1}{a(x_1, x_{14})} + \frac{1}{a(x_{14}, x_{15})}, \tag{14}$$

which is the AC analogue to (4). As a consequence, Γ' is the union of the isolated vertex $\{x_{14}\}$ and the Kron reduction of Γ with respect to $\{x_{14}\}$. Also, $a'(x_1, x_{15})$ is equal to the effective admittance in Γ between x_1 and x_{15}, $a^e(x_1, x_{15})$. Analogously to Example 4, the Laplacian of Γ' is the Dirichlet-to-Neumann map of Γ and $\{x_{14}\}$, which gives us the relationship between potential and injected current at $V \setminus \{x_{14}\} = \{x, z\}$ when the potential is harmonic on x_{14}.

In all the examples, the condition number of the operators $\mathcal{M}_{E(\Gamma),\boldsymbol{u}}$ decreases dramatically in the algorithm, being much lower in the last electrical network fitted than in the first one. This is especially pronounced in Examples 4 and 6, in which the condition number in the first iteration is of the order of magnitude of 10^{15}, and it is 13 orders of magnitude lower in the last network topology obtained.

In some cases, the rms of a network decreases after removing some edges and recovering a network solving Problem 2 with the new set of edges. We conjecture that this phenomenon might be due to the finite precision of the interior-point solver.

As a conclusion, the application of Algorithm 2 to different problems of sparse recovery of an electrical network has yielded promising results. In all cases the algorithm recovers a sparse network after few iterations. In this sparse network topology, the Problem 2 of network recovery becomes well-posed.

Competing Interests The author was funded by a FPI grant of the Research Project PGC2018-096446-B-C21 (with the help of the FEDER Program).

The author was partially supported by the project PID2022-138906NB-C21 funded by MICIU/AEI/ 10.13039/501100011033 and by ERDF/EU. The author was partially supported by the project 2024PROD00092 funded by AGAUR.

The author has no conflicts of interest to declare that are relevant to the content of this chapter.

References

1. Abur, A., Gómez, A.: Power System State Estimation: Theory and Implementation. Marcel Dekker Inc. (2004)
2. Batson, J., Spielman, D., Srivastava, N., Teng, S.: Spectral sparsification of graphs: theory and algorithms. Commun. ACM **56**, 87–94 (2013)
3. Bottura, R., Babazadeh, D., Zhu, K., Borghetti, A., Nordström, L., Nucci, C.A.: Sitl and hla co-simulation platforms: Tools for analysis of the integrated ict and electric power system. In: Eurocon 2013, pp. 918–925. IEEE (2013)
4. Carmona, Á.: Boundary value problems on finite networks. In: Encinas, A.M., Mitjana, M. (eds.) Combinatorial Matrix Theory. Advanced Courses in Mathematics. CRM Barcelona, pp. 173–217. Birkhäuser/Springer, Cham (2018)
5. Chou, H., Maly, J., Verdun, C.: Non-negative Least Squares via Overparametrization (2022). arXiv:2207.08437
6. Deka, D., Backhaus, S., Chertkov, M.: Structure learning in power distribution networks. IEEE Trans. Control Netw. Syst. **5**, 1061–1074 (2015)
7. Fraunhofer IEE and University of Kassel: Pandapower documentation. CIGRE networks. https://pandapower.readthedocs.io/en/v2.6.0/networks/cigre.html. Accessed 22 July 2024
8. Fraunhofer IEE and University of Kassel: Pandapower documentation. Kerber Networks. https://pandapower.readthedocs.io/en/v2.6.0/networks/kerber.html. Accessed: 22 July 2024
9. Gernandt, H., Rohleder, J.: A Calderón type inverse problem for tree graphs. Linear Algebra Appl. **646**, 29–42 (2022)
10. Khan, A., Chakrabarti, S., Sharma, A., Alam, M.: Parameter and topology estimation for electrical power distribution system. In: 2019 8th International Conference on Power Systems (ICPS), pp. 1–5 (2019)
11. Montes, A., Castro, J.: Solving the load flow problem using Gröbner basis. ACM SIGSAM Bull. **29**, 1–13 (1995)
12. Samperio, Á.: Some inverse problems on finite networks. Ph.D Thesis, University of Valladolid (2024)
13. Slawski, M., Hein, M.: Non-negative least squares for high-dimensional linear models: Consistency and sparse recovery without regularization. Electron. J. Statist. **7**, 3004–3056 (2013)
14. Spielman, D.: Graphs, Vectors and Matrices. Bull. Am. Math. Soc. (N.S.) **54**(1), 45–61 (2017)
15. Spielman, D., Srivastava, N.: Graph sparsification by effective resistances. SIAM J. Comput. **40**(6), 1913–1926 (2011)
16. Spielman, D., Teng, S.H.: Spectral sparsification of graphs. SIAM J. Comput. **40**, 981–1025 (2011)
17. Stagg, G., El-Abiad, A.: Computer Methods in Power System Analysis. McGraw-Hill Series in Electronic Systems. McGraw-Hill (1968)
18. Wadhwa, C.L.: Electrical Power Systems. New Academic Science Limited (2012)
19. Webpage: Casadi (2024). https://web.casadi.org/
20. Webpage: Ipopt documentation (2024). https://coin-or.github.io/Ipopt/

A Mathematical Model of Liver Regeneration

Benito Chen-Charpentier

Abstract The liver has a great capacity to regenerate after partial hepatectomy but it is not always successful. Liver cells proliferate to replace the lost metabolic capacity. This paper introduces two mathematical models based on ordinary differential equations for the regeneration of the liver after partial hepatectomy. The cell cycle has three cell types, quiescent, primed and replicating, and the main driver of the cycle is the deficit in the metabolic capacity of the liver. The analysis of the model is limited by the complicated expressions for the steady solutions and their stability. Numerical simulations are done to study different scenarios. A global sensitivity analysis shows that the biggest indices correspond to the failure rates of the process. The paper also gives realizations for different values of the parameters and shows that the regeneration process is successful for many ranges of the parameter values.

Keywords Liver regeneration · Mathematical models · Numerical simulations

1 Introduction

Injuries and diseases are common causes of organ and tissue damage in animals and humans. Many organs and tissues in humans and animals regenerate after an injury that is not very severe. There is a wide body of research on tissue regeneration in animals. See, for example, [23, 60]. In relatively simpler animals like axolotls and zebrafish, regeneration is more prevalent. The second papers argues that evolution works against regeneration. That agrees with the fact that in mammals including humans, tissue regeneration is more limited. For example, if a skin wound is not severe, the skin tissue cells will work together with the immune system and stem cells to repair the damage, [22, 66]. If the wound also involves skeletal muscle and bone damage, the wound can also heal with the immune system and stem cells applying critical roles, [3, 20, 53, 74]. Also smooth muscle tissue can regenerate. The stem cells involved are called called pericytes, and are found in some small blood vessels.

B. Chen-Charpentier (✉)
Department of Mathematics, University of Texas at Arlington, Arlington, TX, USA
e-mail: bmchen@uta.edu

J. R. Torregrosa (ed.), *Mathematical Modelling in Engineering Human Behaviour*, Springer Proceedings in Mathematics & Statistics 513,
https://doi.org/10.1007/978-3-032-00567-0_5

With the participation of pericytes, smooth muscle cells regenerate and repair much more readily than skeletal muscle tissue cells [17, 33]. While in many aspects similar to skeletal muscle tissue, cardiac muscle tissue does not regenerate, or regenerate very little due to the presence of stem cells in the blood that may occasionally enter cardiac tissue. Its regeneration is still a controversial topic [43, 62]. When an injury happens, the body's immune system responds with inflammation. There is a common sequence of steps: chemical signals from injured cells and innate immune cells cause vasodilation; then comes blood clotting; then the complement system promotes inflammation and microbial destruction if necessary; then comes the innate response; and finally the acquired immune response [67, 69]. The replacing of tissue can be done by fibroblasts that deposit collagen, as happens for example in skeletal muscle [53] and in the heart [27], and/or by replacing damaged or dead cells by adult stem cells, as happens in epithelial and connective tissues. It is generally only considered regeneration when damaged or dead cells are replaced by healthy cells. Baddour et al. [5] gives a list of human tissues that regenerate and the processes involved. They include thymus, adrenal gland, thyroid gland, intestine, lungs, liver, blood vessels, eye tissues, kidney and bladder, skin, pancreas, bone, and cartilage. The heart, the brain and other parts of the nervous system do not regenerate. The liver is in a special category since it can regenerate after a loss of about 2/3 its mass [1, 11]. Also, differing from other regeneration processes, the new cells do not come from stem cells but from growth and mitosis of the liver cells. There is a large body of literature on experimental work on liver regeneration [25, 42, 50]. While it is commonly accepted that stem cells play no or little role in the liver regeneration process there is some research showing they do [21, 49]. A partial hepatectomy, or liver resection, is a type of surgery designed to remove cancerous tumors or benign growths from the liver. It is a good alternative for localized, small tumors. But it can also be used to cut lesions from injuries caused by physical objects, such as knives or bullets, or by chemical substances, or radiation. The liver has a great capacity for regeneration if there are no other diseases or factors such as age. These factors can keep the liver from completely regenerating from an up to 70% hepatectomy. Many of the experiments are done with animal models such as mice and rats, but not all the processes are the same as for humans. So there is a need for mathematical models that can help understand the processes involved, try different scenarios and help decide what are the better options for a fast and complete recovery. Existing diseases can hamper the recovery by affecting the signaling chain or keeping the DNA from replicating successfully. The hepatocytes or liver cells reproduce to repair the injury but they also move. In [65] the authors give some references to recent modeling efforts. Existing mathematical models go from the very simple based on ordinary differential equations such as that of [78] that assumes the liver cells grow logistically and [80] that considers hepatocytes and related immune cells grow to occupy the damaged region keeping the same percentages as those in the healthy regions. In [29, 55] are models considering different cells and chemical signals both from cell to cell as well as inside the cells. A more complete model that also considers different species and diseases is in [13]. In this paper we present a model of liver regeneration after resection based on ordinary differential equations.

It considers the same three main cell phases in the cell cycle as the models in [13, 55, 75], but the signaling processes are simplified. The system of equations is solved numerically and comparisons are made with experimental data for mice for a 70% resection. But simulations are also done for other percentages of resection. Since the measured values of the parameters have a large variation, a global sensitivity analysis is performed [63] to determine which parameters have the largest effects on the solutions. The simulations done for different values of the sample space are also solutions of the corresponding random equation model. The rest of the paper is organized as follows: in Sect. 2 the main model of liver resection is presented as well as variations of it. In Sect. 3 the numerical simulations are shown together with the results of the sensitivity analysis. Finally Sect. 4 has discussion and conclusions.

2 Mathematical Models and Methods

In this section we present the regeneration process of the liver after partial hepatectomy. We develop two models of the liver cell cycle responsible for the regeneration and present their equilibrium states and stability. We also so present the global sensitivity analysis used to determine which parameters have the most influence in the changes of the solutions. We also use the simulations necessary for the global sensitivity analysis to obtain realizations of the corresponding random equation model.

2.1 Liver Regeneration

The largest gland in the human body is the liver. It weights about 1.5 kg in an adult. It is an accessory digestive organ producing bile and other chemicals but it also has other essential roles in metabolism, detoxification, synthesis of proteins and production of other substances. These functions make the liver one of the most important organs. Oxygenated blood from the heart goes to the liver The hepatic portal vein delivers partially deoxygenated blood containing nutrients. In addition to nutrients, the liver also absorbs other chemical species including toxins. The liver processes these chemicals and releases them back into the blood for use by other cells [17].

The liver may be damaged by different factors, including injuries caused by external objects, chemicals as well as by many liver diseases and age. There are many kinds of liver diseases: diseases caused by viruses, such as hepatitis A, hepatitis B, and hepatitis C; diseases caused by drugs, poisons, or too much alcohol, examples include fatty liver disease and cirrhosis; liver cancer; and inherited diseases, such as hemochromatosis and Wilson disease. Liver diseases cause more than 3.5% of all deaths worldwide [4, 32].

A partial hepatectomy, or liver resection, is a type of surgery designed to remove cancerous tumors or benign growths from the liver It is also used for lesions caused by injury. It is a good treatment alternative for localized, small tumors, either cancerous, precancerous or benign [44]. The liver has a great capacity for regeneration, especially if there are no other diseases. Liver can even regenerate from a 70% hepatectomy as shown in mouse models [37]. Hepatocytes (parenchymal cells) are the basic cells of the liver, representing 60% of the total cell number and 80% of the total liver volume (Vekemans 2005). They are in charge of the majority of the liver functions such as metabolism and detoxification. The sinusoidal cells are the predominant non-parenchymal cells, forming about 35% of the total cell number and about 17% of the total volume of the liver. These cells include sinusoidal endothelial cells (44%), Kupffer cells (33%), stellate cells (10–25%) and hepatic NK cells (5%). The proportion of the different liver cells does not change with time. In a healthy adult liver hepatocytes are quiescent with only 1 in 20000 cells in a cell cycle [29, 51, 54]. The liver has to satisfy the body metabolic needs. It has the size and the number of hepatocytes to do it. After hepatectomy, there is cell size increase and proliferation to satisfy the metabolic load. If the liver is larger than normal with more hepatocytes than necessary there is apoptosis [10, 26]. The cell cycle is an ordered sequence of events in which the cell grows and then divides into two cells. The cell cycle consists of two main phases: interphase and mitosis. The interphase is subdivided into three sub-phases: G1 in which the cell grows, has biochemical changes and accumulates energy, S or synthesis in which DNA synthesis occurs, and G2 in which more growth occurs. The mitotic phase consists of mitosis in which the DNA is separated and cytokinesis in which the cytoplasm is divided and two daughter cells are produced. The daughter cells may renter the cell cycle if necessary [12, 29]. There are two checkpoints: the G1/S and the G2/M to prevent errors in the replication process [24, 48]. In an otherwise healthy liver, the vast majority of the cells in the cycle complete it successfully, but the presence of other factors such as a disease or age, increase the failure rate [46]. Figure 1 has a sketch of the main phases in the cell cycle.

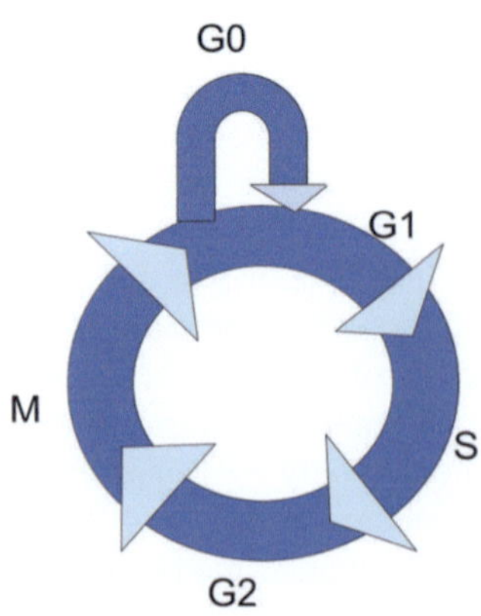

Fig. 1 Sketch of the main phases of the replication cell cycle

2.2 Mathematical Models

We introduce two models of liver regeneration after partial hepatectomy based on ordinary differential equations. The first one is simpler and does not take into account that the cell cycle process may fail. The second one does and is more general. Following [14, 29, 55] we consider three populations of cells: quiescent (Q) for cells before the G1/S checkpoint, primed (P) for cells between the G1/S and G2/M checkpoints and replicating (R) after. Q cells get primed by a deficit in the metabolic capacity, growth factors and excess immediate genes. The P cells move into R cells due to growth factors and metabolic deficit. There is also a chain of cytokines inside the hepatocytes that result in the desired growth and duplication. Le Roy et al. [45] has a list of growth factors and cytokines involved and in [76] there is a review of mathematical models for the cell cycle that use different hypotheses for the cytokines, growth factors and other signals involved. Other papers describing the factors involved in the cell cycle are [8, 30, 51]. For simplicity we will assume that the deficit in the metabolic capacity of the liver is the main reason for the priming of the Q cells and for the transition from P cells to R cells. We assume that the time that the signaling chain and the processes take is included in the rate constants, k_Q, k_P and k_R. We normalize the number of cells in a healthy adult liver to be one, and assume that they have the capacity to keep the metabolic processes going. We define $N = Q + G_P P + G_R R$ and the metabolic deficit as $N - 1$ since for a healthy liver $N = Q = 1$ and the deficit is zero. G_P and G_R are the increment in size factors of P and R cells respectively. We assume that there are both hypertrophy or cell size increment accounted by the G increment factor; and hyperplasia or cell proliferation [14, 76]. The metabolic deficit is $k(1 - N)$, with k a positive constant. Once the metabolic deficit is zero, the cell cycle stops. In the first model we take $G_P = G_R = G$. We also assume that there is no apoptosis [19, 38, 52]. There are no checkpoints, so once the cycle starts it always ends successfully in two new cells This is the situation for an otherwise healthy young adult liver [36, 72]. The cell cycle process ends once the liver can satisfy the metabolic needs, that is when $N = 1$. The first model is given by the following set of ordinary differential equations, with all parameters positive

$$
\begin{aligned}
N &= Q + G(P + R) \\
\frac{d}{dt}Q &= -k_Q\,(1 - N)\,Q + 2k_R R \\
\frac{d}{dt}P &= k_Q\,(1 - N)\,Q - k_P\,(1 - N)\,P \\
\frac{d}{dt}R &= k_P\,(1 - N)\,P - k_R R.
\end{aligned}
\tag{1}
$$

Theorem 1 *For $Q, P, R \in [0, 1]$, the system of Eqs.* (1) *has three equilibrium solutions with $R = 0$: $Q = 0, P = 0, R = 0$; $Q = 1 - GP, R = 0$; $Q = \frac{k_P}{Gk_Qk_P}$, $P = \frac{k_Q}{Gk_Qk_P}$, $R = 0$. Since the algebraic sytem of equations for the equilibrium solutions is* *formed by three quadratic equations, there are other solutions with no simple closed form.*

Proof Letting $R = 0$, it is straight forward to find a line segment of solutions $Q = 1 - QP$, $P \in [0, 1/G]$, $Q \in [0, 1]$. The third solution is found using Mathematica [39].

To determine the local stability of the equilibrium solutions it is necessary to calculate the eigenvalues of the matrix of the corresponding linearized system about the given equilibrium solution. This requires solving a cubic algebraic equation whose solutions are complicated expressions with few exceptions. One exception is linearizing about $Q = 1$, $P = 0$, $R = 0$ which gives the eigenvalues $\lambda = 0, -k_R, k_Q(1 - G)$. The third eigenvalue is non-positive since $G \geq 1$. But the presence of the zero eigenvalue requires to expand the central manifold about the equilibrium point [56]. The result of this laborious process is that $Q = 1$, $P = 0$, $R = 0$ is locally unstable. For the other equilibrium points te process yields very complicated formulas. But it can be shown that the zero equilibrium $Q = 0$, $P = 0$, $R = 0$ is locally unstable using the following argument: start from $Q = q$, $P = 0$, $R = 0$ with q 'small'. Since the metabolic load is not satisfied ($N << 1$), at least one quiescent cell enters into the cell cycle and exits as two daughter cells. So the cell population increases and the zero equilibrium is locally unstable.

Model (1) will be used to do regeneration simulations for different percentages of the partial hepatectomy.

The second model is an improvement of the first since it adds that the cells start growing before the first checkpoint and finish growing before the second checkpoint [9, 28]. Therefore $G_P < G_R$. It also considers that a certain percentage of the cells p fail the first checkpoint, and a percentage $p1$ the second one. This may be due to the presence of diseases or to aging [36, 72]. The cells that fail are removed from the system since they have some damage.

$$
\begin{aligned}
N =& Q + G(P + R)\\
\frac{d}{dt}Q =& -k_Q\,(1 - N)\,Q + 2k_R R\\
\frac{d}{dt}P =& k_Q\,(1 - N)\,Q(1 - p) - k_P\,(1 - N)\,P\\
\frac{d}{dt}R =& k_P(1 - p1)\,(1 - N)\,P - k_R R.
\end{aligned}
\tag{2}
$$

Theorem 2 *System* (2) *with* $p = p1 = 0$ *has the same equilibrium points with* $R = 0$ *than system* (1).

The proof is immediate. But the local stability formulas are more complicated. The exception is that the equilibrium $Q = 1$, $P = 0$, $R = 0$ has the same eigenvalues as for (1) and is locally stable from the analysis of the central manifold. In the next section we will do numerical simulations calculating solutions that go to equilibrium states. Since the numerical solutions have truncation and round-off errors, the calculated equilibrium solutions are locally stable.

We have already mentioned the papers with models of liver regeneration [14, 29, 55] that have the same cells but add explicitly the effect of liver Kuppfer and Stellate cells, as well as signaling inside the hepatocytes. Other papers are [78] where only the hepatocytes are considered and they have a growth rate that is a product of logistics. The paper [80] considers hepatocytes, Kuffer cells, macrophages, and endothelial cells and they grow keeping the same percentages. Banks and Thompson [6] considers the number of times a cell goes into the cycle. Yates et al. [79] considers the cell cycle as a Markov chain. There are also models based on partial differential equations, for example, [35].

2.3 Sensitivity Analysis

The mathematical models presented depend on various parameters. The first one given by (1) depends on k_Q, k_P, k_R and G. The second one (2) depends on G_P and G_R instead of on G. It also depends on p and $p1$. Determining the values of these parameters is difficult. The parameters k_Q, k_P and k_R depend not only on the time the cells spend in each phase but also on the metabolic load. The percentage of cells that fail to replicate is not easy to measure The size increment parameters G_P and G_R can be measured but the cell populations have large variations. There are also uncertainties and errors associated with their estimation, many times by fitting the solutions to measured data. But their variability also causes uncertainties in the solutions of the models. Therefore it is important to determine how sensitive the outputs of the model are to variations in each of the parameter values. If small changes in the value of a parameter causes large changes in the solution, there is a large uncertainty in the in the output. It may also indicate that there are errors, or that the model is incorrect. Very small changes in the solutions of the model when changing the value of a parameter may indicate that the terms involving the that parameter may be eliminated from the model that can thus be simplified. There are two main types of sensitivity analysis, local and global. A local sensitivity analysis is appropriate if the parameter values are known accurately and there are no interactions between the different parameters [2, 18, 34, 47]. This is not usually the situation in models related to Biology that involve processes with large variability, uncertainty and possible errors. So a global sensitivity analysis (GSA) is more adequate since it analyzes investigates the sensitivity over the entire space of possible parameter values, and includes the possible interactions between the effects of the different parameters [61, 63]. Global sensitivity analysis is more expensive but also more reliable. There are different global sensitivity methods. The most commonly used are sobol [63, 64, 81], FAST [15, 63], and EFAST [16, 57]. To reduce the number of sensitivity indices we will only consider the effect of the parameter changes on $N = Q + G_P P + G_R R$, which is directly related to the metabolic capacity.. We will also take the average, maximum and minimum values of the sensitivity indices over the time interval of interest.

In global sensitivity analysis methods, sample values of the parameters are taken over their ranges of interest and, for models based in differential equations, the model is solved over the time range of interest. This is equivalent of solving the corresponding random model with the parameters having probabilistic distributions. Therefore we also have simulations including randomness, as we are solving the corresponding random differential equations model [70]. Thus we can incorporate uncertainty into the model results without having to introduce stochastic differential equations [41] or using the stochastic simulation algorithm [31]. We will use Sobol's method to calculate the global sensitivity indices. It is variance-based [63].

3 Results

In this section we will show the results for simulations based on data for mice for 70% hepatectomy. While there is data also for rats for 70% hepatectomy, we will use the data for mice used in Fig. 9 of Cook's paper [14, 68]. Cook and Vadigepalli [14] also presents results for rats based on data from [73] and humans [59]. We will only do simulations for mice since there are more data points and we concentrate on different scenarios such as varying the percentage of the hepatectomy and changing the effect of the checkpoints.

3.1 Simulations

We start by fitting the parameters k_q, k_P and k_R to the experimental data [14, 68] with $G_P = 1.75$, $G_R = 2.0$, $p = 0$, $p1 = 0$. These values correspond to a young healthy liver with growth of 75% before the synthesis phase and 25% before the mitotic phase. In the literature, for example [40, 77], the values are not given but these numbers agree with the descriptions of the relative sizes and that replicating cells divide into two normal size daughter cells. The values of the fitted rate parameters are in Table 1.

Figure 2 has the plots of the cells over time for Model (2) using the rate parameter values given in Table 1, and $G_P = 1.75$, $G_R = 2.$, $p = 0$, $p1 = 0$. The simulations correspond to a 70% hepatectomy. The liver regenerates. The steady state

Table 1 Values of the rate parameters in the models

Parameter	Value
k_Q	1.05e-1
k_P	6.6e-2
k_R	6.e-2

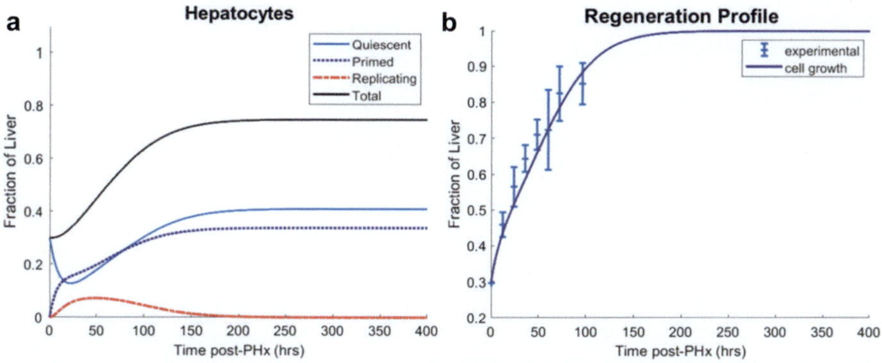

Fig. 2 Model (2) using the rate parameter values given in Table 1, $G_P = 1.75, G_R = 2, p = 0, p1 = 0$ for a 70% hepatectomy. **a** Cell populations. **b** Metabolic capacity N and experimental data

is $Q = .399$, $P = .337$, $R = 0$. 70% hepatectomy is the most common one done in animal experiments. Since G_P, G_R, p, $p1$ are given and we are only fitting the rate parameters k_Q, k_P and k_R the fit is not very good but still reasonable, especially considering the size of the error bars. We will keep these values of the rate parameters in all the simulations. The regeneration profile refers to N.

Figure 3 has the plots of the cells over time for Model (2) using the rate parameter values given in Table 1, and $G_P = 1.75$, $G_R = 2.$, $p = 0$, $p1 = 0$, corresponding to a 30% hepatectomy. The liver regenerates faster compared to the 70% hepatectomy. Also the steady state is $Q = .479$, $P = .297$, $R = 0$. 30% hepatectomy is the second most common experimental liver resection procedure.

Figure 4 has the plots of the cells over time for Model (2) using the rate parameter values given in Table 1, and $G_P = 1.75$, $G_R = 2.$, $p = 0$, $p1 = 0$, corresponding to a 10% hepatectomy. The liver regenerates faster compared to the 30% hepatectomy.

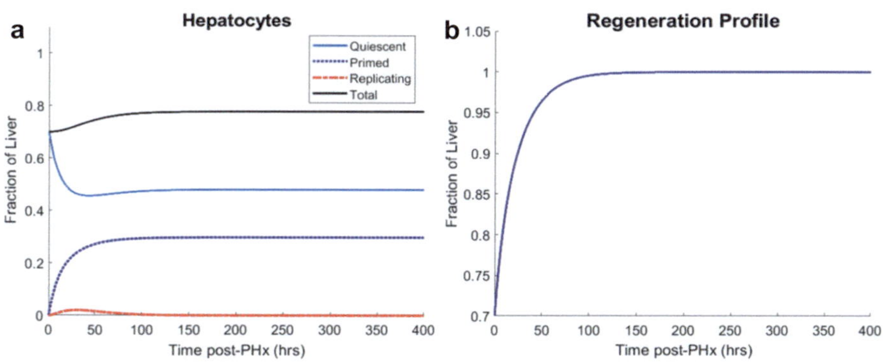

Fig. 3 Model (2) using the rate parameter values given in Table 1, $G_P = 1.75, G_R = 2, p = 0, p1 = 0$ for a 30% hepatectomy. **a** Cell populations. **b** Metabolic capacity N

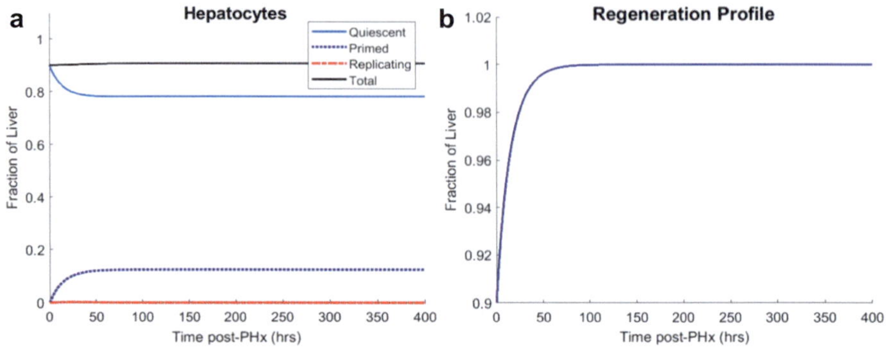

Fig. 4 Model (2) using the rate parameter values given in Table 1, $G_P = 1.75$, $G_R = 2$, $p = 0$, $p1 = 0$ for a 10% hepatectomy. **a** Cell populations. **b** Metabolic capacity N.

Also the steady solution has $R = 0$ and the percentage of P cells is smaller than for the two previous figures.

Next we assume that all the growth happens in the P cells, and $G_P = G_R = 2$. We show the plots for a 70% hepatectomy in Fig. 5. Not much change compared with Fig. 2 but the number of P cells is smaller.

The corresponding plots for a 30% hepatectomy in Fig. 6. Not much change compared with Fig. 2 but the number of P cells is smaller.

In the regeneration models in [29, 55] the assumption is that there is no hypertrophy. That is, $G_P = G_R = 1$. We show the plots for a 70% hepatectomy in Fig. 7. Now the number of Q and P cells is larger compared with Fig. 2. The total number of cells is equal to N.

For a 30% hepatectomy with $G_P = G_R = 1$. the plots are in Fig. 8. Now the number of Q and P cells is larger compared with Fig. 3.

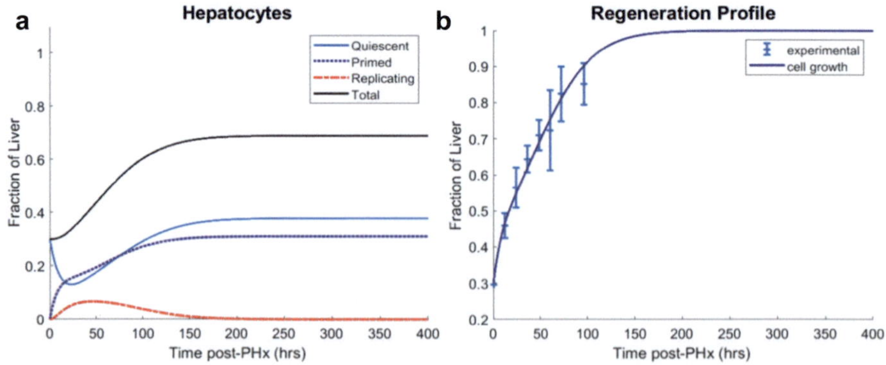

Fig. 5 Model (2) using the rate parameter values given in Table 1, $G_P = G_R = 2$, $p = 0$, $p1 = 0$ for a 70% hepatectomy. **a** Cell populations. **b** Metabolic capacity N and experimental data

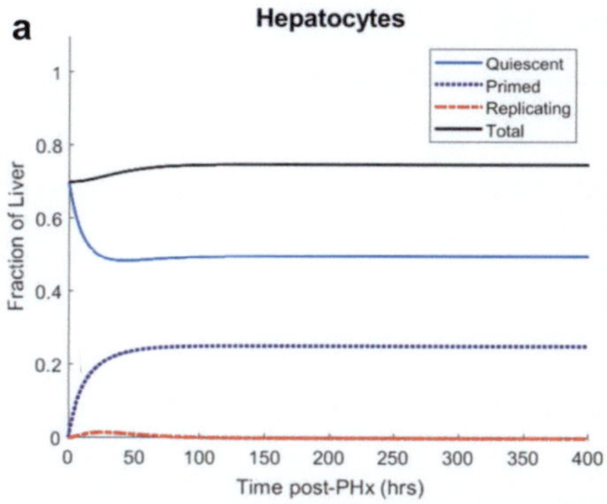

Fig. 6 Model (2) using the rate parameter values given in Table 1, $G_P = G_R = 2$, $p = 0$, $p1 = 0$ for a 30% hepatectomy. **a** Cell populations. **b** Metabolic capacity N

Fig. 7 Model (2) using the rate parameter values given in Table 1, $G_P = G_R = 1$, $p = 0$, $p1 = 0$ for a 70% hepatectomy. **a** Cell populations

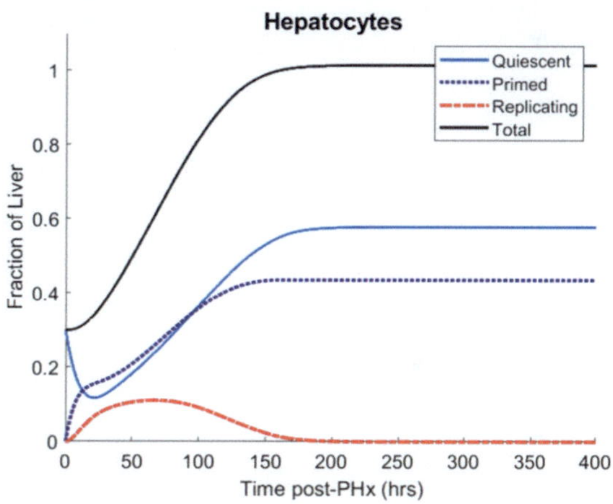

Now we consider that not all cells entering the cell cycle are successful in replicating. The failure percentage at the first check point is p and at the second $p1$. We start by taking $p = p1$ and increasing their value until the regeneration process fails. This happens for both 70% and 30% hepatectomies for values of $p = p1$ between .4 and .5, as shown in Figs. 9, 10, 11 and 12. Figures 9 and 10 are for $p = p1 = .4$ and the liver still regenerates but it takes a much longer time. Figures 11, 12 are for $p = p1 = .5$ and the liver only partially regenerates.

To find the relative importance of the two checkpoints, we gave them different values between 0.4 and 0.5. Figures 13 and 14 show that for $p = 0.5$ and $p1 = 0.4$ the liver regenerates for both a 70% and a 30% hepatectomy. But Figs. 15 and 16 show that for $p = 0.4$ and $p1 = 0.5$ the liver fails to completely regenerate for both a 70% and a 30% hepatectomy. The effect of the second checkpoint is more important. This result agrees with the results of the sensitivity analysis presented in the next subsection

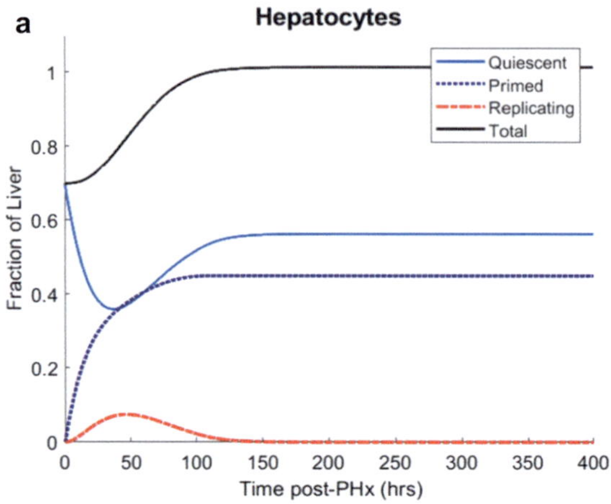

Fig. 8 Model (2) using the rate parameter values given in Table 1, $G_P = G_R = 1.$, $p = 0$, $p1 = 0$ for a 30% hepatectomy. **a** Cell populations

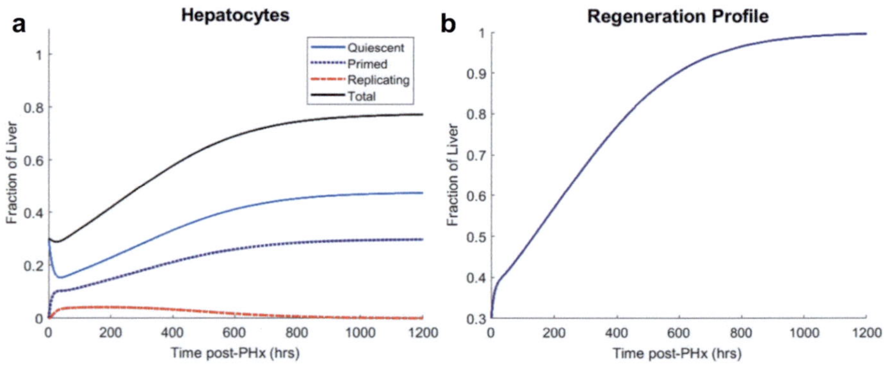

Fig. 9 Model (2) using the rate parameter values given in Table 1, $G_P = 1.75$, $G_R = 2.0$, $p = 0.4$, $p1 = 0.4$ for a 70% hepatectomy. **a** Cell populations. **b** Metabolic capacity N

Finally we will assume that the presence of other diseases and/or aging causes the cell cycle to take more time but that there are no checkpoints. That is, $p = p1 = 0$ and the values of k_Q, k_P and k_R from Table 1 are divided by a factor of 1.2. The results for a 70% hepatectomy are shown in Fig. 17 and compared with those in Fig. 2 show only a slowing of the processes.

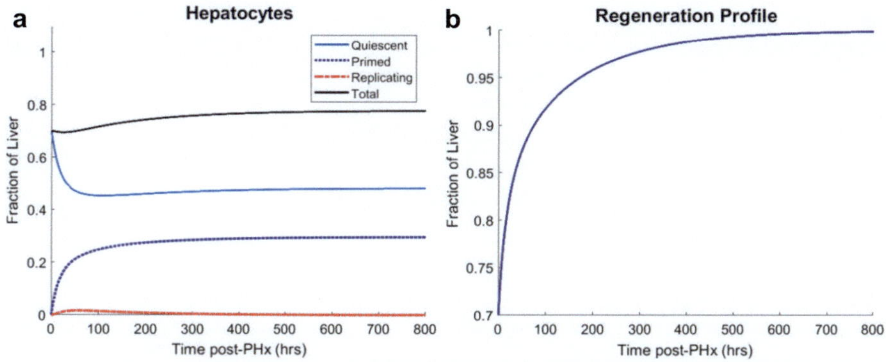

Fig. 10 Model (2) using the rate parameter values given in Table 1, $G_P = 1.75$, $G_R = 2.0$, $p = 0.4$, $p1 = 0.4$ for a 30% hepatectomy. **a** Cell populations. **b** Metabolic capacity N

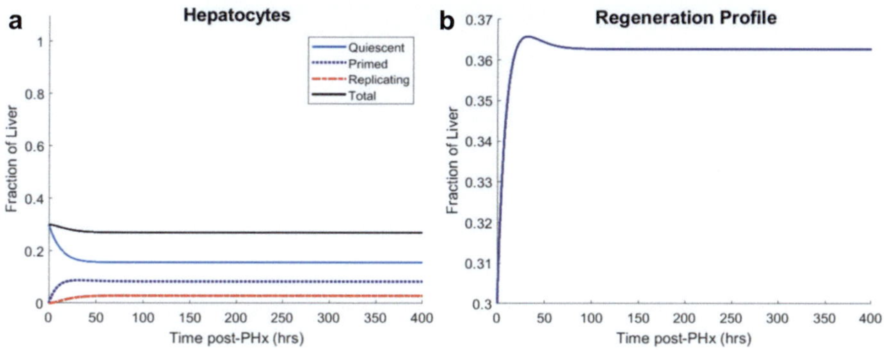

Fig. 11 Model (2) using the rate parameter values given in Table 1, $G_P = 1.75$, $G_R = 2.0$, $p = 0.5$, $p1 = 0.5$ for a 70% hepatectomy. **a** Cell populations. **b** Metabolic capacity N

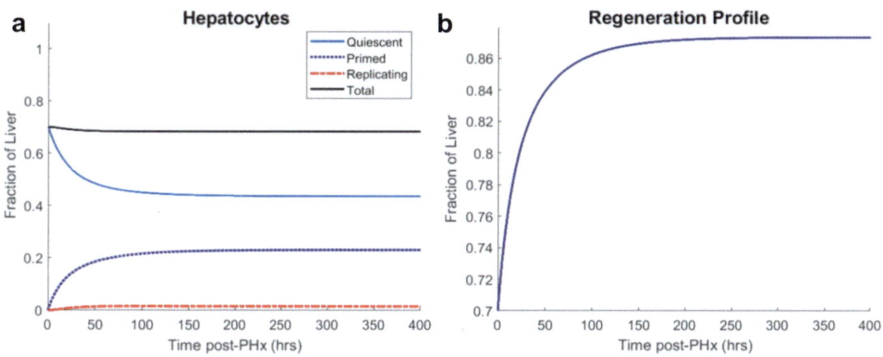

Fig. 12 Model (2) using the rate parameter values given in Table 1, $G_P = 1.75$, $G_R = 2.0$, $p = 0.5$, $p1 = 0.5$ for a 30% hepatectomy. **a** Cell populations. **b** Metabolic capacity N

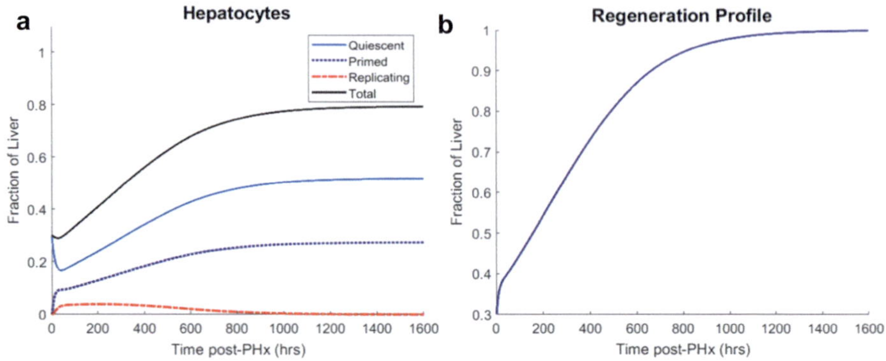

Fig. 13 Model (2) using the rate parameter values given in Table 1, $G_P = 1.75$, $G_R = 2.0$, $p = 0.5$, $p1 = 0.4$ for a 70% hepatectomy. **a** Cell populations. **b** Metabolic capacity N

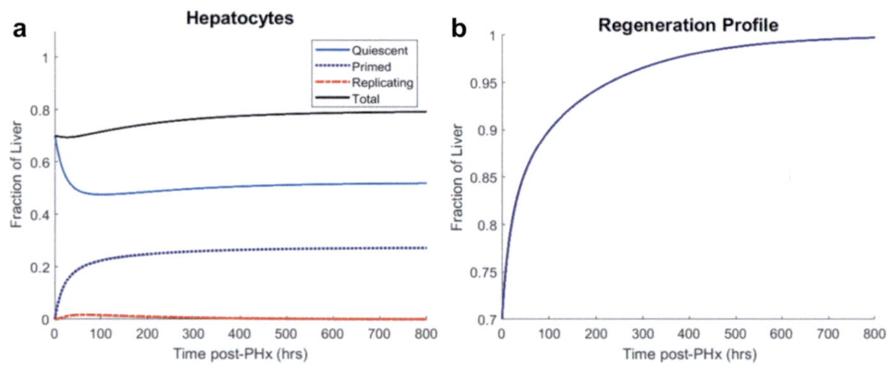

Fig. 14 Model (2) using the rate parameter values given in Table 1, $G_P = 1.75$, $G_R = 2.0$, $p = 0.5$, $p1 = 0.4$ for a 30% hepatectomy. **a** Cell populations. **b** Metabolic capacity N

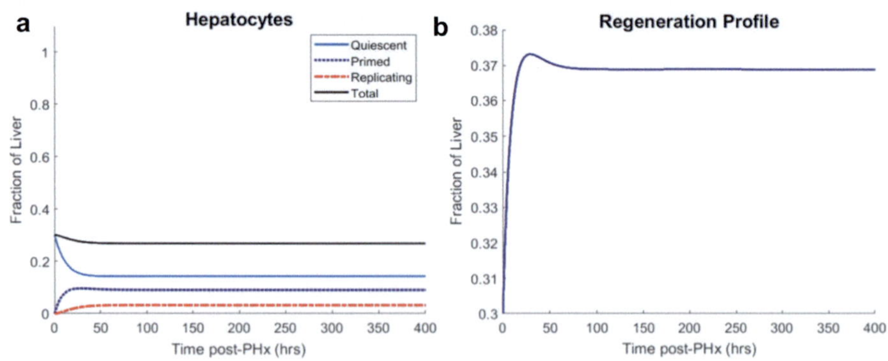

Fig. 15 Model (2) using the rate parameter values given in Table 1, $G_P = 1.75$, $G_R = 2.0$, $p = 0.4$, $p1 = 0.5$ for a 70% hepatectomy. **a** Cell populations. **b** Metabolic capacity N

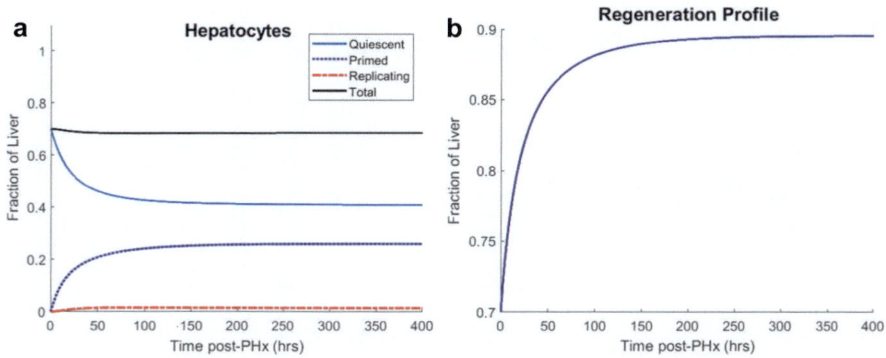

Fig. 16 Model (2) using the rate parameter values given in Table 1, $G_P = 1.75$, $G_R = 2.0$, $p = 0.4$, $p1 = 0.5$ for a 30% hepatectomy. **a** Cell populations. **b** Metabolic capacity N

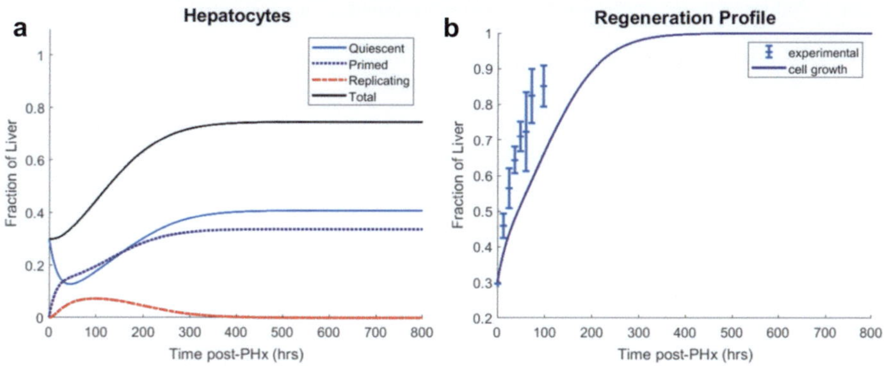

Fig. 17 Model (2) using the rate parameter values given in Table 1 divided by 1.2, $G_P = 1.75$, $G_R = 2.0$, $p = 0.0$, $p1 = 0.0$ for a 70% hepatectomy. **a** Cell populations. **b** Metabolic capacity N and experimental data

3.2 *Sensitivity Analysis*

For the sensitivity analysis we used the SAFE Toolbox [57, 58]. It runs in matlab/octave and calculates the global sensitivity indices for models based on ordinary differential equations. We used Sobol's method [63] which is variance based. We found the indices over the whole recovery interval (400 h), and Fig. 18 shows the average, maximum and minimum values of the metabolic capacity N with respect to all parameters. The average index is the most significant. The main effect indices are considering that there is no interaction between the parameters, and the total effect indices are when there is. The total effect indices are more informative. Since we have no information on the distributions we assumed that they were are uniform. The distributions used are in Table 2. The left plot in the figures is for a 70% and the right one for a 30% hepatectomy. The largest sensitivity index corresponds to $p1$.

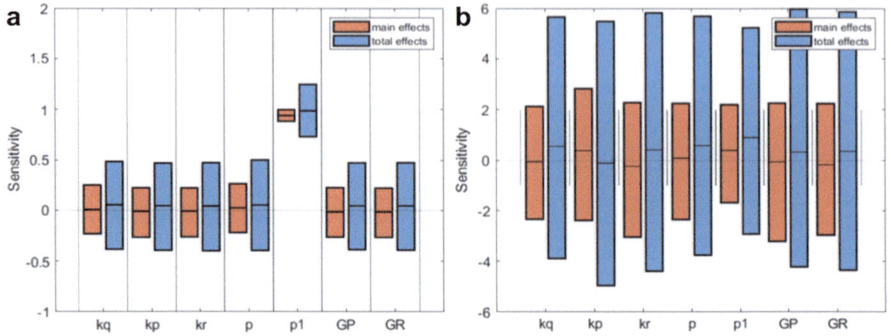

Fig. 18 Sobol's indices of N in Model (2) using the distributions in Table 2. **a** 70% hepatectomy. **b** 30% hepatectomy

Table 2 Values of the uniform distributions of the parameters in the models

Parameter	Distribution
k_Q	[0.5e-1, 1.5e-1]
$k)P$	[3.3e-2, 9.9e-2]
k_R	[3e-3, 9e-2]
G_P	[1.5, 1.8]
G_R	[1.8, 2.1]
p	[0, 0.5]
$p1$	[0, 0.5]

The effect of $p1$ is much more noticeable for a 70% hepatectomy since the liver has to regenerate more.

The calculation of the global sensitivity indices involves solving the model at sample values of the parameters. If we consider that the model (2) is a random differential equations model with random coefficients with uniform distributions, we already have realizations of the solution. Figures 19 and 20 show in the left plot all the calculated trajectories for N. There are many trajectories for which the liver regenerates completely but also trajectories for which it does not. The right plot shows the average of all the trajectories, and the average +/– one standard deviation. This shows that even with fairly large variations of the parameters, the model shows that the liver regenerates. Figure 19 if for a 70% hepatectomy and Fig. 20 for a 30% one.

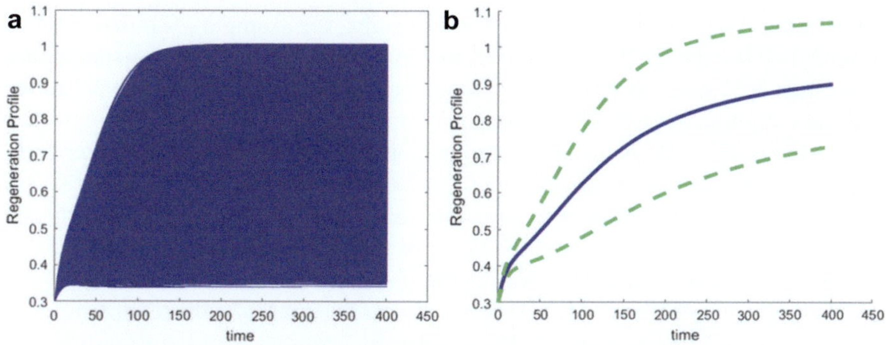

Fig. 19 Trajectories of N using the distributions in Table 2 for 70% hepatectomy. **a** All trajectories. **b** Average, and average +/– one standard deviation

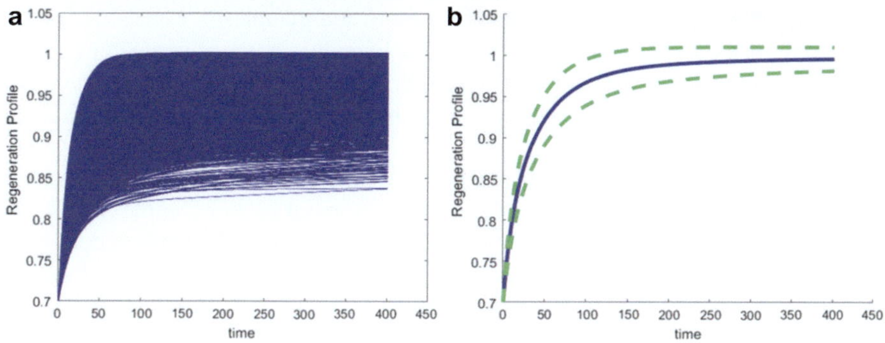

Fig. 20 Trajectories of N using the distributions in Table 2 for 30% hepatectomy. **a** All trajectories. **b** Average, and average +/– one standard deviation.

4 Conclusions

We introduced two new models for liver regeneration after partial hepatectomy using ordinary differential equations in three cell phases: quiescent, primed and replicating. We assumed that cell replication is controlled by the metabolic capacity deficiency, and the time rates of change also include the total time for the chemical signaling and replication processes to happen. The first one (1) is simpler and consider that the total growth in size happens to the primed cells, and that once a cell enters the cell cycle it succeeds in replicating. The second model (2) adds growth for both the primed and replicating cells and the two fail safe checkpoints. For the first model (1) some analysis can be done but since the right hand side of the differential equations are quadratic in the unknowns, most of the results have to be done numerically. Even more for the second model. The first simulations were done to find the effect of the percentage of the hepatectomy on the generation. As expected as the percentage increases, the time for regeneration increases but the liver can still regenerate. The regeneration times agree with experimental results, for example [77], that say that for rodents it is between 7 an 10 days. Simulations were not done for hepatectomies higher than 70% since for those the metabolic capacity of the liver will be below the minimum required by the body, and the individual will die. The proportion of primed cells to quiescent cells increases as the percentage of the hepatectomy decreases, mostly in agreement with experimental results given, for example, in [77], that for 30% hepatectomy there is only hypertrophy, and for 70% there is hypertrophy and proliferation. Next we considered the effect of the amount of hypertrophy (cell size increase) and whether it happens only before the first checkpoint or also after it. For a cell to replicate into two normal size cells, the size factor G_R of the replicating cells should be about 2. The duration of the regeneration process depends slightly on the values of G_p and G_R. If $G_P = G_R = 1$ then the process is much slower. The biggest changes in the regeneration process come from the percentage of cells that fail. The failure can be due to breaks in the signaling chain or to errors in the duplication of the DNA, or of the nucleus, or in the division of the cytoplasm. For values of the failing rates at each checkpoint about 50%, the liver does not regenerate completely, with the failure rate at the second checkpoint having the bigger effect. Therefore the health and age of the patient should be strong consideration for performing a hepatectomy.

Some regeneration processes are still unclear. Mathematical models can help. But since there are not many empirical data, models cannot be too complex. As more data is available, the models should consider more processes. Also, cancer cells have cell cycles without check points. There is current research in trying to add them [7, 71] and mathematical models can help.

Competing Interests The author has no conflicts of interest to declare that are relevant to the content of this chapter.

References

1. Andersen, K.J., Knudsen, A.R., Kannerup, A.-S., Sasanuma, H., Nyengaard, J.R., Hamilton-Dutoit, S., Erlandsen, E.J., Jørgensen, B., Mortensen, F.V.: The natural history of liver regeneration in rats: description of an animal model for liver regeneration studies. Int. J. Surg. **11**(9), 903–908 (2013)
2. Arriola, L., Hyman, J.M.: Sensitivity analysis for uncertainty quantification in mathematical models. In: Mathematical and Statistical Estimation Approaches in Epidemiology, pp. 195–247. Springer (2009)
3. Arvidson, K., Abdallah, B., Applegate, L., Baldini, N., Cenni, E., Gomez-Barrena, E., Granchi, D., Kassem, M., Konttinen, Y., Mustafa, K., et al.: Bone regeneration and stem cells. J. Cell Mol. Med. **15**(4), 718–746 (2011)
4. Asrani, S.K., Devarbhavi, H., Eaton, J., Kamath, P.S.: Burden of liver diseases in the world. J. Hepatol. **70**(1), 151–171 (2019)
5. Baddour, J.A., Sousounis, K., Tsonis, P.A.: Organ repair and regeneration: an overview. Birth Defects Res. C. Embryo Today **96**(1), 1–29 (2012)
6. Banks, H.T., Thompson, W.C.: Mathematical models of dividing cell populations: application to cfse data. Math. Model. Nat. Phenomena **7**(5), 24–52 (2012)

7. Bisteau, X., Caldez, M.J., Kaldis, P.: The complex relationship between liver cancer and the cell cycle: a story of multiple regulations. Cancers **6**(1), 79–111 (2014)
8. Caldas, H., Jiang, Y., Holloway, M.P., Fangusaro, J., Mahotka, C., Conway, E.M., Altura, R.A.: Survivin splice variants regulate the balance between proliferation and cell death. Oncogene **24**(12), 1994–2007 (2005)
9. Caldez, M.J., Van Hul, N., Koh, H.W., Teo, X.Q., Fan, J.J., Tan, P.Y., Dewhurst, M.R., Too, P.G., Talib, S.Z.A., Chiang, B.E., et al.: Metabolic remodeling during liver regeneration. Dev. Cell **47**(4), 425–438 (2018)
10. Christ, B., Collatz, M., Dahmen, U., Herrmann, K.-H., Höpfl, S., König, M., Lambers, L., Marz, M., Meyer, D., Radde, N., et al.: Hepatectomy-induced alterations in hepatic perfusion and function-toward multi-scale computational modeling for a better prediction of post-hepatectomy liver function. Front. Physiol. **12**, 733868 (2021)
11. Christ, B., Dahmen, U., Herrmann, K.-H., König, M., Reichenbach, J.R., Ricken, T., Schleicher, J., Ole Schwen, L., Vlaic, S., Waschinsky, N.: Computational modeling in liver surgery. Front. Physiol. **8** (2017)
12. Clark, M.A., Douglas, M., Choi, J.: Biology 2e. openstax (2020)
13. Cook, D., Ogunnaike, B.A., Vadigepalli, R.: Systems analysis of non-parenchymal cell modulation of liver repair across multiple regeneration modes. BMC Syst. Biol. **9**(1), 1–24 (2015)
14. Cook, D., Vadigepalli, R.: Computational modeling as an approach to study the cellular and molecular regulatory networks driving liver regeneration. In: Liver Regeneration, pp. 185–198. Elsevier (2015)
15. Cukier, R., Fortuin, C., Shuler, K.E., Petschek, A., Schaibly, J.H.: Study of the sensitivity of coupled reaction systems to uncertainties in rate coefficients. i theory. J. Chem. Phys. **59**(8), 3873–3878 (1973)
16. Dela, A., Shtylla, B., de Pillis, L.: Multi-method global sensitivity analysis of mathematical models. J. Theor. Biol. **546**, 111159 (2022)
17. DeSaix, P., Betts, G.J., Johnson, E., Johnson, J.E., Oksana, K., Kruse, D.H., Poe, B., Wise, J.A., Young, K.A.: Anatomy & Physiology (openstax) (2013)
18. Dickinson, R.P., Gelinas, R.J.: Sensitivity analysis of ordinary differential equation systems—a direct method. J. Comput. Phys. **21**(2), 123–143 (1976)
19. Ding, B.-S., Nolan, D.J., Butler, J.M., James, D., Babazadeh, A.O., Rosenwaks, Z., Mittal, V., Kobayashi, H., Shido, K., Lyden, D., et al.: Inductive angiocrine signals from sinusoidal endothelium are required for liver regeneration. Nature **468**(7321), 310–315 (2010)
20. Duda, G.N., Geissler, S., Checa, S., Tsitsilonis, S., Petersen, A., Schmidt-Bleek, K.: The decisive early phase of bone regeneration. Nat. Rev. Rheumatol. **19**(2), 78–95 (2023)
21. Duncan, A.W., Dorrell, C., Grompe, M.: Stem cells and liver regeneration. Gastroenterology **137**(2), 466–481 (2009)
22. El-Ashram, S., El-Samad, L.M., Basha, A.A., El Wakil, A.: Naturally-derived targeted therapy for wound healing: beyond classical strategies. Pharmacol. Res. **170**, 105749 (2021)
23. Elchaninov, A., Sukhikh, G., Fatkhudinov, T.: Evolution of regeneration in animals: a tangled story. Front. Ecol. Evol. **9**, 621686 (2021)
24. Elledge, S.J.: Cell cycle checkpoints: preventing an identity crisis. Science **274**(5293), 1664–1672 (1996)
25. Fausto, N.: Liver regeneration. J. Hepatol. **32**, 19–31 (2000)
26. Francavilla, A., Ove, P., Polimeno, L., Coetzee, M., Makowka, L., Barone, M., Van Thiel, D., Starzl, T.: Regulation of liver size and regeneration: importance in liver transplantation. In: Transplantation Proceedings, vol. 20, p. 494. NIH Public Access (1988)
27. Frangogiannis, N.G.: Inflammation in cardiac injury, repair and regeneration. Curr. Opin. Cardiol. **30**(3), 240–245 (2015)
28. Fuβ, H., Dubitzky, W., Downes, C.S., Kurth, M.J.: Mathematical models of cell cycle regulation. Brief. Bioinf. **6**(2), 163–177 (2005)
29. Furchtgott, L.A., Chow, C.C., Periwal, V.: A model of liver regeneration. Biophys. J . **96**(10), 3926–3935 (2009)

30. Gilgenkrantz, H., de l'Hortet, A.C.: Understanding liver regeneration: from mechanisms to regenerative medicine. Am. J. Pathol. **188**(6), 1316–1327 (2018)
31. Gillespie, D.T.: Stochastic simulation of chemical kinetics. Annu. Rev. Phys. Chem. **58**(1), 35–55 (2007)
32. Gong, J., Tu, W., Liu, J., Tian, D.: Hepatocytes: a key role in liver inflammation. Front. Immunol. **13**, 1083780 (2023)
33. Grounds, M.D., White, J.D., Rosenthal, N., Bogoyevitch, M.A.: The role of stem cells in skeletal and cardiac muscle repair. J. Histochem. Cytochem. **50**(5), 589–610 (2002)
34. Gustafson, P., Srinivasan, C., Wasserman, L.: Local sensitivity analysis. Bayesian Stat. **5**, 197–210 (1996)
35. Hoehme, S., Bertaux, F., Weens, W., Grasl-Kraupp, B., Hengstler, J.G., Drasdo, D.: Model prediction and validation of an order mechanism controlling the spatiotemporal phenotype of early hepatocellular carcinoma. Bull. Math. Biol. **80**, 1134–1171 (2018)
36. Hora, S., Wuestefeld, T.: Liver injury and regeneration: current understanding, new approaches, and future perspectives. Cells **12**(17), 2129 (2023)
37. Hori, T., Ohashi, N., Chen, F., Baine, A.-M.T., Gardner, L.B., Jermanus, S., Nguyen, J.H.: Simple and sure methodology for massive hepatectomy in the mouse. Ann. Gastroenterol. Q. Publ. Hell. Soc. Gastroenterol. **24**(4), 307 (2011)
38. Iimuro, Y., Nishiura, T., Hellerbrand, C., Behrns, K.E., Schoonhoven, R., Grisham, J.W., Brenner, D.A., et al.: Nfkappab prevents apoptosis and liver dysfunction during liver regeneration. J. Clin. Investig. **101**(4), 802–811 (1998)
39. Inc., W.R.: Mathematica, Version 14.1. Champaign, IL (2024)
40. Kholodenko, I.V., Yarygin, K.N.: Cellular mechanisms of liver regeneration and cell-based therapies of liver diseases. Biomed. Res. Int. **2017**(1), 8910821 (2017)
41. Kloeden, P., Platen, E.: Applications of mathematics. In: Numerical Solution of Stochastic Differential Equations, vol 23. Springer (1992)
42. Koniaris, L.G., McKillop, I.H., Schwartz, S.I., Zimmers, T.A.: Liver regeneration. J. Am. Coll. Surg. **197**(4), 634–659 (2003)
43. Laflamme, M.A., Murry, C.E.: Heart regeneration. Nature **473**(7347), 326–335 (2011)
44. Lau, W.: Future perspectives for hepatocellular carcinoma. Hpb **5**(4), 206–213 (2003)
45. Le Roy, B., Dupré, A., Gallon, A., Chabrot, P., Gagnière, J., Buc, E.: Liver hypertrophy: underlying mechanisms and promoting procedures before major hepatectomy. J. Visc. Surg. **155**(5), 393–401 (2018)
46. Ma, C., Gurkan-Cavusoglu, E.: A comprehensive review of computational cell cycle models in guiding cancer treatment strategies. NPJ Syst. Biol. Appl. **10**(1), 71 (2024)
47. Ma, Y., Dixit, V., Innes, M.J., Guo, X., Rackauckas, C.: A comparison of automatic differentiation and continuous sensitivity analysis for derivatives of differential equation solutions. In: 2021 IEEE High Performance Extreme Computing Conference (HPEC), pp. 1–9. IEEE (2021)
48. Matthews, H.K., Bertoli, C., de Bruin, R.A.: Cell cycle control in cancer. Nat. Rev. Mol. Cell Biol. **23**(1), 74–88 (2022)
49. Matthews, V.B., Yeoh, G.C.: Liver stem cells. IUBMB Life **57**(8), 549–553 (2005)
50. Michalopoulos, G.K.: Liver regeneration. The Liver: Biology and Pathobiology, pp. 566–584 (2020)
51. Michalopoulos, G.K., Bhushan, B.: Liver regeneration: biological and pathological mechanisms and implications. Nat. Rev. Gastroenterol. Hepatol. **18**(1), 40–55 (2021)
52. Miyaoka, Y., Ebato, K., Kato, H., Arakawa, S., Shimizu, S., Miyajima, A.: Hypertrophy and unconventional cell division of hepatocytes underlie liver regeneration. Curr. Biol. **22**(13), 1166–1175 (2012)
53. Musarò, A.: The basis of muscle regeneration. Adv. Biol. **2014**(1), 612471 (2014)
54. Ozaki, M.: Cellular and molecular mechanisms of liver regeneration: proliferation, growth, death and protection of hepatocytes. In: Seminars in Cell & Developmental Biology, vol. 100, pp. 62–73. Elsevier (2020)
55. Periwal, V., Gaillard, J., Needleman, L., Doria, C.: Mathematical model of liver regeneration in human live donors. J. Cell. Physiol. **229**(5), 599–606 (2014)

56. Perko, L.: Differential Equations and Dynamical Systems, vol. 7. Springer Science & Business Media (2013)
57. Pianosi, F., Sarrazin, F., Wagener, T.: A matlab toolbox for global sensitivity analysis. Environ. Model. Softw. **70**, 80–85 (2015)
58. Pianosi, F., Sarrazin, F., Wagener, T.: How successfully is open-source research software adopted? Results and implications of surveying the users of a sensitivity analysis toolbox. Environ. Model. Softw. **124**, 104579 (2020)
59. Pomfret, E.A., Pomposelli, J.J., Gordon, F.D., Erbay, N., Price, L.L., Lewis, W.D., Jenkins, R.L.: Liver regeneration and surgical outcome in donors of right-lobe liver grafts. Transplantation **76**(1), 5–10 (2003)
60. Poss, K.D.: Advances in understanding tissue regenerative capacity and mechanisms in animals. Nat. Rev. Genet. **11**(10), 710–722 (2010)
61. Rabitz, H.: General sensitivity analysis of differential equation systems. In: Fluctuations and Sensitivity in Nonequilibrium Systems: Proceedings of an International Conference, University of Texas, Austin, Texas, 12–16 Mar 1984, pp. 196–203. Springer (1984)
62. Sadek, H., Olson, E.N.: Toward the goal of human heart regeneration. Cell Stem Cell **26**(1), 7–16 (2020)
63. Saltelli, A., Ratto, M., Andres, T., Campolongo, F., Cariboni, J., Gatelli, D., Saisana, M., Tarantola, S.: Global Sensitivity Analysis: The Primer. John Wiley & Sons (2008)
64. Savatorova, V.: Exploring parameter sensitivity analysis in mathematical modeling with ordinary differential equations. CODEE J. **16**(1), 4 (2023)
65. Seddek, A.-L., Hassan, R.: Modelling of liver regeneration after hepatectomy. Arch. Toxicol. **94**(10), 3605–3606 (2020)
66. Sen, C.K.: MicroRNA in Regenerative Medicine. Academic Press (2023)
67. Sherwood, E.R., Toliver-Kinsky, T.: Mechanisms of the inflammatory response. Best Pract. Res. Clin. Anaesthesiol. **18**(3), 385–405 (2004)
68. Shu, R.-Z., Zhang, F., Wang, F., Feng, D.-C., Li, X.-H., Ren, W.-H., Wu, X.-L., Yang, X., Liao, X.-D., Huang, L., et al.: Adiponectin deficiency impairs liver regeneration through attenuating stat3 phosphorylation in mice. Lab. Invest. **89**(9), 1043–1052 (2009)
69. Smith, C., Kruger, M.J., Smith, R.M., Myburgh, K.H.: The inflammatory response to skeletal muscle injury: illuminating complexities. Sports Med. **38**, 947–969 (2008)
70. Soong, T.T., Bogdanoff, J.: Random differential equations in science and engineering
71. Sung, W., Hong, T.S., Poznansky, M.C., Paganetti, H., Grassberger, C.: Mathematical modeling to simulate the effect of adding radiation therapy to immunotherapy and application to hepatocellular carcinoma. Int. J. Radiat. Oncol. Biol. Phys. **112**(4), 1055–1062 (2022)
72. Tanaka, M., Miyajima, A.: Liver regeneration and fibrosis after inflammation. Inflamm. Regen. **36**, 1–6 (2016)
73. Tanoue, S., Uto, H., Kumamoto, R., Arima, S., Hashimoto, S., Nasu, Y., Takami, Y., Moriuchi, A., Sakiyama, T., Oketani, M., et al.: Liver regeneration after partial hepatectomy in rat is more impaired in a steatotic liver induced by dietary fructose compared to dietary fat. Biochem. Biophys. Res. Commun. **407**(1), 163–168 (2011)
74. Tidball, J.G., Villalta, S.A.: Regulatory interactions between muscle and the immune system during muscle regeneration. Am. J. Physiol. Regul. Integr. Comparat. Physiol. **298**(5), R1173–R1187 (2010)
75. Verma, B.K., Subramaniam, P., Vadigepalli, R.: Model-based virtual patient analysis of human liver regeneration predicts critical perioperative factors controlling the dynamic mode of response to resection. BMC Syst. Biol. **13**, 1–15 (2019)
76. Williams, K.S., Secomb, T.W., El-Kareh, A.W.: An autonomous mathematical model for the mammalian cell cycle. J. Theor. Biol. **569**, 111533 (2023)
77. Yagi, S., Hirata, M., Miyachi, Y., Uemoto, S.: Liver regeneration after hepatectomy and partial liver transplantation. Int. J. Mol. Sci. **21**(21), 8414 (2020)
78. Yamamoto, K., Ishii, M., Inoue, Y., Hirokawa, F., MacArthur, B., Nakamura, A., Haeno, H., Uchiyama, K.: Prediction of postoperative liver regeneration from clinical information using a data-led mathematical model. Sci. Rep. **6** (2016)

79. Yates, C.A., Ford, M.J., Mort, R.L.: A multi-stage representation of cell proliferation as a Markov process. Bull. Math. Biol. **79**, 2905–2928 (2017)
80. Zafarnia, S., Mrugalla, A., Rix, A., Doleschel, D., Gremse, F., Wolf, S.D., Buyel, J.F., Albrecht, U., Bode, J.G., Kiessling, F., et al.: Non-invasive imaging and modeling of liver regeneration after partial hepatectomy. Front. Physiol. **10**, 904 (2019)
81. Zhang, X.-Y., Trame, M.N., Lesko, L.J., Schmidt, S.: Sobol sensitivity analysis: a tool to guide the development and evaluation of systems pharmacology models. CPT: Pharmacometrics & Systems Pharmacology **4**(2), 69–79 (2015)

A Proposal of a Combined Choquet Index Asssociated with Two Vectors of Weights

José Carlos R. Alcantud

Abstract The topic of combining the ordered weighted average (for short, OWA) operators and weighted average means (for short, WAM) has been extensively researched. This approach has become a prominent method for extending OWA operators. When using mixed indices that incorporate traits from both WAMs and OWAs, two independent sets of weights are required. The weights in WAM are value-focused, while those in OWA are order-focused. This work explores how two independent weights of vectors can be merged to create a single operator that integrates characteristics from both WAMs and, to a lesser extent, OWAs. The resulting operator is expressed as a Choquet integral defined by a 2-additive capacity. In this way, the proposed index uses capacities without requiring further additional information.

Keywords Choquet integral · Fuzzy measure · Aggregation operator · Index

1 Introduction

AOs (which holds for aggregation operators) transform a vector of numerical values into a single value. The OWA operators (which holds for ordered weighted averaging operators) are a class of AOs that are defined from a vector of weights. Since their introduction by Yager [30], refinements have proliferated. Remarkably, these enhancements often use a second vector of weights. We will also be inspired by this approach that resorts to two weighting vectors. An early inspiration is the successful class of WOWA operators [4, 27] defined by Torra [24]. These AOs combined the WAM (which holds for weighted average mean) with OWAs, which are symmetric. WAOWAs were also defined in the continuous case [26]. Applications to models of soft computing are available too [3]. Many other scholars put forward joint operators that merge both approaches. A shortlist of works includes Broomell and Wagner [6], Llamazares [14], and Merigò [19].

J. C. R. Alcantud (✉)
Unidad de Excelencia Gestión Económica para la Sostenibilidad GECOS and IME, University of Salamanca, Salamanca, Spain
e-mail: jcr@usal.es

J. R. Torregrosa (ed.), *Mathematical Modelling in Engineering Human Behaviour*, Springer Proceedings in Mathematics & Statistics 513,
https://doi.org/10.1007/978-3-032-00567-0_6

It is interesting to note that Choquet integrals generalize both OWAs and WAMs [9]. They have been applied in fields including bibliometrics [28], social choice [1], welfare economics [10], machine learning [22], or soft computing [2]. This class of AOs requires the idea of a capacity or fuzzy measure [20]. Capacities are set functions that satisfy monotonicity (extending a subset produces a weakly better evaluation) and their evaluation of the empty set is zero [23]. Informally, capacities are more general than measures because they use monotonicity instead of the stronger additivity property. A problem of this concept is that it requires an exponential number of figures to be completely defined.

WAMs and OWAs are the particular cases of the Choquet integral that are defined by an additive or a symmetric capacity, respectively. This characteristic has led to other Choquet-inspired proposals that generalize both WAMs with OWAs. For example, Yager and Alajlan [31, Sect. 4] proposed to use both capacities, an aggregation attitudinal function plus a quasi-arithmetic generating function, as a way to extend importance weighted mean aggregation. Llamazares [15–17] suggested that using semi-uninorms together with two weighting vectors gives rise to an interesting class of AOs called SUOWAs. They are Choquet integrals defined from capacities that utilize semi-uninorms plus the values of capacities corresponding to OWAs and WAMs. Although WOWAs are a Choquet integral, they are different from SUOWAs [15]. Llamazares [18] emphasized that Choquet integrals are able to generalize both OWAs and WAMs through capacities constructed from two weighting vectors and suitable functions.

This chapter explains in an explicit manner that it is possible to construct a Choquet integral from two vectors of weights, with the help of a suitably defined 2-additive fuzzy measure. This is the simplest non-additive instance of the class of k-additive fuzzy measures defined by Grabisch [11]. The main motivation for these fuzzy measures is reducing the computational complexity of a fuzzy measure, in such way that enough flexibility is retained. The restriction to this notable type of capacities allows us to bypass the need for other external elements, such as those mentioned above, This comes at the cost of restricting the applicability to pairs of vectors that meet adequate requirements.

Our strategy is as follows. We define an explicit set function from the two weighting vectors that are given. Then we check whether the set function behaves as the Möbius inverse of a 2-additive fuzzy measure. In the cases when this is true, then the capacity produces a discrete Choquet integral. This AO is called the joint Choquet index defined from the two weighting vectors.

The key issue in this strategy is to decide when the set function that is proposed in this chapter is the Möbius inverse of a (2-additive) fuzzy measure. Chateauneuf and Jaffray [8] proposed a characterization that solves the general version of the problem (i.e., without any restriction by k-additivity). With the help of the version of their theorem that embeds the 2-additivity property, we are able to propose sufficient conditions for the two weighting vectors to yield a joint Choquet index. We illustrate the use of this technique with fully developed examples. We also study the properties of the new index and expose some limitations of our approach. Particularly, we show

how in certain cases, the index can ve decomposed as the sum of a WAM and a (modified) OWA on a restricted domain of vectors.

We conclude that Choquet integrals associated with 2-additive capacities allow us to generalize WAMs with the incorporation of an independent vector of weights. This definition uses a reduced number of parameters ($2n$ in a case with n components), but it can leverage the crucial Choquet AO with the benefits of two unrelated weighting vectors.

The outline of this chapter is as follows. In Sect. 2 we give background. Then Sect. 3 proposes the joint Choquet index associated with two independent vectors. Some of its fundamental properties are given too. Section 4 investigates cases where this joint index is well-defined. Section 5 gives an example where we compute both the joint Choquet index and comparable indices defined in the literature from two vectors of weights. We give concluding remarks in Sect. 6.

2 Background

The basic setup of our problem depends on the following elements.

Assumptions
Hereafter we shall assume that the set $X = \{1, \ldots, n\}$ represents a finite list of indicators.

Also, we let $\mathbf{w} = (w_1, \ldots, w_n) \in [0, 1]^n$ and $\mathbf{v} = (v_1, \ldots, v_n) \in [0, 1]^n$ be weighting vectors, therefore we assume $\sum_{i=1}^n w_i = \sum_{i=1}^n v_i = 1$.

The goal of this research is to produce evaluations of numerical values indexed by the indicators in X, that incorporate both vectors and have properties inspired by those of the next two aggregation procedures:

Definition 1 The weighted average mean (WAM) associated with $\mathbf{w}$ is the function $WAM^{\mathbf{w}} : \mathbb{R}^n \longrightarrow \mathbb{R}$ defined by the following expression:

$$WAM^{\mathbf{w}}(a_1, \ldots, a_n) = \sum_{i=1}^{n} w_i\, a_i, \quad \text{for every } \mathbf{a} = (a_1, \ldots, a_n) \in \mathbb{R}^n. \tag{1}$$

Definition 2 (Yager [30]) The ordered weighted averaging (OWA) operator associated with $\mathbf{v}$ is the function $O^{\mathbf{v}} : \mathbb{R}^n \longrightarrow \mathbb{R}$ defined by the following expression:

$$O^{\mathbf{v}}(a_1, \ldots, a_n) = \sum_{i=1}^{n} v_i\, b_i, \quad \text{for every } \mathbf{a} = (a_1, \ldots, a_n) \in \mathbb{R}^n. \tag{2}$$

In this expression, b_i is the i-th largest element in the collection of (possibly repeated) values $\{a_1, \ldots, a_n\}$.

This means that when $\mathbf{a}_{\searrow} = (a_{[1]}, \ldots, a_{[n]})$ is a non-increasing permutation of $\mathbf{a}$, then we set $b_i = a_{[i]}$.

The next section recalls fundamental facts about an important aggregation operator that extends the two definitions given above. This operator requires a great deal of information in comparison with the operators considered above, in the form of a concept called capacity. Thus prior to the definition of this aggregation operator (namely, the discrete Choquet integral), we dwell on the properties of capacities and especially, we explore a popular case that simplifies many computations while keeping enough flexibility.

2.1 Choquet Integrals and Related Results

Very important for our exercise is the next type of set-valued functions, that generalize the concept of a measure by weakening the additivity requirement to only monotonicity:

Definition 3 A discrete capacity on X is a set function $\mu : 2^X \longrightarrow [0, 1]$ that is monotonic (i.e., $\mu(A) \leqslant \mu(B)$ whenever $A \subseteq B \subseteq X$) and satisfies $\mu(\varnothing) = 0$, $\mu(X) = 1$.

Sometimes the restriction $\mu(X) = 1$ is dropped, and when it is kept, it is interpreted as a normalization condition.

With respect to terminology, note that although the term 'capacity' was coined by Choquet [9], Sugeno [23] independently defined the same notion under the name fuzzy measure.

A (discrete) capacity μ is additive when in case that $A, B \subseteq X$ are disjoint, then the equality $\mu(A \cup B) = \mu(A) + \mu(B)$ holds true. Observe that additive capacities are measures. Besides, μ is symmetric when $\mu(A) = \mu(B)$ provided that $A, B \subseteq X$ and $|A| = |B|$.

It has been argued that "[p]erhaps the most successful subfamily [of capacities] is the subfamily of k-additive measures, and inside this subfamily, the most appealing case is the case of 2-additive measures" (Miranda and García-Segador [21]). The main reason is computational tractability. To define a capacity on X we need $2^n - 2$ numbers that in addition, must satisfy the mandatory monotonicity constraints. Many researchers have produced simplified but flexible expressions of a capacity, and to this purpose the k-additive capacities have proven to be especially reliable. With $\sum_{i=1}^{k} C_n^i$ evaluations one can to define a k-additive capacity [13]. This concept can be given a compact definition with the help of the Möbius inverse [11], an operator that returns the capacity that produced it through the Zeta transform. Let us define the operator that we need in this work:

Definition 4 ([12, Def. 2.30]) The Möbius inverse of the discrete capacity μ is the unique solution to the following system of equations:

$$\mu(X') = \sum_{A \subseteq X'} m^{\mu}(A), \text{ for each } X' \subseteq X. \quad (3)$$

The solution is the next set function m^{μ} defined on the power set of X:

$$m^{\mu}(X') = \sum_{A \subseteq X'} (-1)^{|X' \setminus A|} \mu(A), \text{ for each } X' \subseteq X.$$

Equation (3) defines the Zeta transform of m^{μ}.

Additive capacities can be characterized as the capacities that have Möbius inverses whose evaluations on sets with two or more elements are zero [12, Th. 2.33]. By inspiration of this characterization, the next properties that extend the idea of additivity have been proposed:

Definition 5 (Grabisch [11]) A discrete capacity μ on X is k-additive (with $k = 1, 2, \ldots$) when a subset of X whose cardinality is k exists, whose evaluation by m^{μ} is not null; and for all $X' \subseteq X$ with at least $k + 1$ elements, the evaluation of X' by m^{μ} is null.

Of course, additive capacities are 1-additive capacities by the aforementioned [12, Th 2.33].

It is important to note that any 2-additive capacity is totally determined by the values that it attains on singletons and pairs of elements, because the following convenient expression can be proven [11, Eq. (5)]

$$\mu(X') = \sum_{\substack{i \neq j \\ i, j \in X'}} \mu(\{i, j\}) - (|X'| - 2) \sum_{i \in X'} \mu(\{i\}), \text{ for each } X' \subseteq X.$$

From this expression, some simple computations demonstrate that $\frac{n(n+1)}{2}$ values suffice to define a 2-additive capacity on a set with n elements. This figure is much smaller than the $2^n - 2$ values that are needed to determine a general capacity (on a set with n elements).

When m^{μ} is the Möbius inverse of a 2-additive capacity μ, it is well known that the following equality must be true:

$$1 - \sum_{i \in X} \mu(\{i\}) = \sum_{\substack{i \neq j \\ i, j \in X}} m^{\mu}(\{i, j\}).$$

The following natural question arises: how can we characterize if a mapping $m : 2^X \longrightarrow \mathbb{R}$ is a 2-additive capacity on X? The next theorem gives an exact solution to this problem:

Theorem 1 (Chateauneuf and Jaffray [8]) *Let $m : 2^X \longrightarrow \mathbb{R}$ be a set-valued function. Then m is the Möbius inverse of a capacity μ on X if and only if the next two restrictions are met:*

(1) $m(\varnothing) = 0$ *and* $\sum_{X' \subseteq X} m(X') = 1$.
(2) When $X' \subseteq X$ *and* $j \in X'$: $\sum_{j \in A \subseteq X'} m(A) \geqslant 0$.

Suppose in particular that we have defined a set-valued function m on 2^X for which we know that $m(X') = 0$ whenever $|X'| > 2$. In case that we need to use Chateauneuf and Jaffray's test to guarantee that m is a 2-additive capacity, it is important to state that the second condition in Theorem 1 reduces to the following property:

(3) *When* $X' \subseteq X$ *and* $j \in X'$: $\sum_{i \in X'} m(\{i, j\}) \geqslant 0$.

The latter condition imposes $m(\{i, j\}) + \mu(\{j\}) \geqslant 0$ for all i, j (which is the application of the property to the case $X' = \{i, j\}$ and j). The restriction enforces monotonicity of μ defined from m by the Zeta transform, whereas (1) assures the so-called boundary conditions (i.e., $\mu(\varnothing) = 0$ and $\mu(X) = 1$).

We are ready to recall that discrete capacities produce discrete Choquet integrals by the following rule:

Definition 6 (Choquet [9]) The discrete Choquet integral with respect to the capacity μ is $C^\mu : \mathbb{R}^n_+ \longrightarrow \mathbb{R}$ where $C^\mu(a_1, \ldots, a_n) = \sum_{j=1}^n \left[a_{(j)} - a_{(j-1)}\right] \mu(L_j)$, and $\mathbf{a}_\nearrow = (a_{(1)}, \ldots, a_{(n)})$ is a non-decreasing permutation of $\mathbf{a} = (a_1, \ldots, a_n)$, we let $a_{(0)} = 0$, and $L_j = \{(j), \ldots, (n)\}$ is the set of indices of the $n - j + 1$ largest components of $\mathbf{a}$.

Important
It is remarkable that there is an explicit correspondence between WAMs (resp., OWAs) and Choquet integrals defined from additive (resp., symmetric) capacities (v., [12, Theorem 4.63], also [17, Remark 3]).

Especially for our needs, the WAM defined from $\mathbf{w}$ on $\mathbb{R}^n_+$ is the Choquet integral with respect to the additive capacity $\mu^{\mathbf{w}}$ such that $\mu^{\mathbf{w}}(\{i\}) = w_i$ for all $i \in X$.

2.2 Extensions of WAM and OWA

Several attempts have been made to combine the WAM and OWA respectively defined from two weighting vectors. It has been argued that they extend the original operators in different forms. Examples of composed operators include the following list:

1. The ordered weighted averaging-weighted average (abbreviated, OWAWA) operator [19] utilizes a convex combination of two aggregation operators: with the help of an extra parameter $\alpha \in [0, 1]$, that does not depend on the other inputs, we let

$$OWAWA(\mathbf{a}) = \alpha WAM^{\mathbf{w}}(\mathbf{a}) + (1 - \alpha) O^{\mathbf{v}}(\mathbf{a}), \quad \text{for every } \mathbf{a} \in \mathbb{R}^n. \tag{4}$$

2. The standard deviation OWA (abbreviated, SDOWA) operator [7] utilizes the same formula, but in this case the proponents insist that the α weighting parameter should be directly defined from $\mathbf{w}$ and $\mathbf{v}$. And they propose that this should be done by their standard deviations using the next expression:

$$\bar{\alpha} = \frac{sd(\mathbf{w})}{sd(\mathbf{w}) + sd(\mathbf{v})}.$$

   This choice guarantees the coincidence with $O^{\mathbf{v}}$, the OWA operator with vector $\mathbf{v}$. when $\mathbf{w}$ is constant. It also guarantees that SDOWA coincides with $WAM^{\mathbf{w}}$ when $\mathbf{v}$ is constant.
   Anyhow, the expression that calculates the value of SDOWA applied to $\mathbf{a}$ becomes

$$\bar{\alpha} WAM^{\mathbf{w}}(\mathbf{a}) + (1 - \bar{\alpha}) O^{\mathbf{v}}(\mathbf{a}).$$

3. The HWA (hybrid weighted average) operator [29] composes the OWA defined from $\mathbf{w}$ with a function directly defined from the other vector $\mathbf{v}$. Specifically, the application on $\mathbf{a} \in \mathbb{R}^n$ is

$$HWA(\mathbf{a}) = O^{\mathbf{v}}(n\mathbf{wa}) = O^{\mathbf{v}}(nw_1a_1, \ldots, nw_na_n).$$

4. The JWA (joint weighted average) operator [6] allows the weights to interact. As in the case of SDOWA, it does not introduce further parameters.
   In order to apply JWA on a vector $\mathbf{a} \in \mathbb{R}^n$, we need to use $\mathbf{a}_{\searrow} = (a_{[1]}, \ldots, a_{[n]})$, which is a non-increasing permutation of $\mathbf{a}$, and then we must produce the similarly ordered vector of weights $\mathbf{w}_{[\mathbf{a}]} = (w_{[1]}, \ldots, w_{[n]})$, Then we should apply the formula

$$JWA(\mathbf{a}) = (\mathbf{w}_{[\mathbf{a}]} \oplus \mathbf{v})\mathbf{a}_{\searrow} \tag{5}$$

   where the definition of $\oplus$ is given by compositional geometry as follows:

$$\mathbf{x} \oplus \mathbf{y} = \left(\frac{x_1 y_1}{\sum_{i=1}^n x_i y_i}, \ldots, \frac{x_n y_n}{\sum_{i=1}^n x_i y_i} \right) \text{ if } \mathbf{x} = (x_1, \ldots, x_n),\ \mathbf{y} = (y_1, \ldots, y_n).$$

We do not recall WOWAs here. It is appropriate to explain that this AO uses an interpolation function, in addition to the two vectors of weights. in addition, we recall that the result depends on the choice of the interpolation function, which is not uniquely determined.

3 Main Concept and its Basic Properties

Within the study of aggregation operators that use **w** and **v**, we proceed to investigate the properties of a proposal that requires the introduction of neither further parameters nor external functions. By doing so we avoid the problem of eliciting the corresponding additional factor.

3.1 The Joint Choquet Index: Definition and Basic Properties

We begin by producing a set function as follows:

Definition 7 Let $P^{\mathbf{w},\mathbf{v}} = \frac{\sum_{i \neq j}(w_i w_j - v_i v_j)}{\binom{n}{2}}$.

Define the following set function $m^{\mathbf{w},\mathbf{v}}$ on X:

$$m^{\mathbf{w},\mathbf{v}}(X') = \begin{cases} 0, & \text{when } X' = \varnothing \text{ or } X' \subseteq X, |X'| > 2 \\ w_i, & \text{when } X' = \{i\} \text{ for some } i \in X, \\ w_i w_j - v_i v_j - P^{\mathbf{w},\mathbf{v}}, & \text{when } X' = \{i, j\} \text{ for some } i, j \in X, \ i \neq j. \end{cases}$$

This auxiliary function leads us to the main definition that we propose in this work:

Definition 8 In case that $m^{\mathbf{w},\mathbf{v}}$ produces a capacity $\mu^{\mathbf{w},\mathbf{v}}$, then we say that $C^{\mu^{\mathbf{w},\mathbf{v}}}$, i.e., the Choquet integral defined by $\mu^{\mathbf{w},\mathbf{v}}$, is the joint Choquet index associated with **w** and **v**.

Note that by definition, the joint Choquet index defined above shares the following property with the WAM defined from **w**: both indices are Choquet integrals respectively defined by capacities $\mu^{\mathbf{w},\mathbf{w}}$ and $\mu^{\mathbf{w}}$ such that $\mu^{\mathbf{w},\mathbf{w}}(\{i\}) = \mu^{\mathbf{w}}(\{i\}) = w_i$ for all $i \in X$.

Question

The concept given in Definition 8 raises a natural question that we investigate below: what properties of the weighting vectors produce a joint Choquet index associated with them?

The next section expands on the properties of the joint Choquet index associated with identical vectors, and other remarkable cases. Afterwards we perform a study on the conditions on **w** and **v** that produce a well-defined joint Choquet index in Sect. 4.

3.2 Properties of the Joint Choquet Index

We proceed to state several specific properties of the joint Choquet index given in Definition 8. The standard properties derived from the fact that Definition 8 is constructed as a Choquet integral must hold true too. Therefore the joint Choquet index is compensative, monotonic, idempotent, and satisfies positive homogeneity of degree 1.

The properties below will help us identify similarities (and differences) with WAMs and their hybridizations with OWAs.

Note first that the construction in Definition 7 guarantees that whenever $\mu^{\mathbf{w},\mathbf{v}}$ is a capacity, then it must be either additive or 2-additive. Below we find several cases where we have identifiable additive capacities, which necessarily must produce the WAM associated with **w**.

The proof of most of the following properties are routine. Below we delve into the arguments proving the last two claims, that are more technical.

1. For each $i \in X$, $\mu^{\mathbf{w},\mathbf{v}}(\{i\}) = m^{\mathbf{w},\mathbf{v}}(\{i\}) = w_i$.
   This is a compulsory assignment for capacities producing additive Choquet integrals that coincide with the WAM that **w** defines [12, Eq. (4.77)].
2. When $\mathbf{w} = \mathbf{v}$, the joint Choquet index associated with **w** and **v** is well defined and it coincides with $WAM^{\mathbf{w}} = WAM^{\mathbf{v}}$.
   Note that $P^{\mathbf{w},\mathbf{v}} = 0$, and in fact, when $i \neq j$ it is the case that $m^{\mathbf{w},\mathbf{v}}(\{i, j\}) = 0$, which proves additivity.
3. In the more particular case where $\mathbf{w} = \mathbf{v} = (\frac{1}{n}, \ldots, \frac{1}{n})$, the joint Choquet index associated with **w** and **v** coincides with the simple average mean (i.e., the arithmetic average).
   This is the only OWA operator defined as a joint Choquet index associated with two vectors using Definition 8.
4. When **v** is such that $v_i = 1$ for some i, then $\mu^{\mathbf{w},\mathbf{v}}$ is additive if and only if it is symmetric.
   And in this case, $\mathbf{w} = (\frac{1}{n}, \ldots, \frac{1}{n})$, therefore the joint Choquet index associated with these vectors is the simple arithmetic average (i.e, the WAM associated with equal weights $\mathbf{w} = (\frac{1}{n}, \ldots, \frac{1}{n})$).
5. When **w** is such that $w_i = 1$ for some i, then the only capacity $\mu^{\mathbf{w},\mathbf{v}}$ that is additive is the Dirac measure centered at i.
   It is defined by

$$\mu^{\mathbf{w},\mathbf{v}}(X') = \begin{cases} 1, & \text{when } i \in X', \\ 0, & \text{otherwise} \end{cases} \quad \text{(cf., [12, Sect. 2.2])}$$

exactly through the vector of weights $\mathbf{v} = (\frac{1}{n}, \ldots, \frac{1}{n})$.

In this case, [12, Cor. 4.64 (iii)] assures that the joint Choquet index associated with these vectors is the projection on the i-th coordinate. Of course, it is the WAM associated with $\mathbf{w}$.

6. Suppose $\mathbf{w} = \left(\frac{1}{n}, \ldots, \frac{1}{n}\right)$, and there are $i_0 \neq j_0$ such that $v_{i_0} + v_{j_0} = 1$. Then the joint Choquet index associated with $\mathbf{w}$ and $\mathbf{v}$ is well defined. It attains the value
$$WAM^{\mathbf{w}}(\mathbf{a}) + v_{i_0} \cdot (1 - v_{i_0}) \cdot OWA^{\mathbf{n}}(\mathbf{a}) \text{ with } \mathbf{n} = \left(0, \frac{1}{\binom{n}{2}}, \ldots, \frac{n-2}{\binom{n}{2}}, \frac{n-1}{\binom{n}{2}}\right)$$
at every $\mathbf{a} \in \mathbb{R}^n_+$ with $a_{i_0} \cdot a_{j_0} = 0$.
7. Suppose $\mathbf{v} = \left(\frac{1}{n}, \ldots, \frac{1}{n}\right)$, and there are $i_0 \neq j_0$ such that $w_{i_0} + w_{j_0} = 1$. Then the joint Choquet index associated with $\mathbf{w}$ and $\mathbf{v}$ is well defined. It attains the value
$$WAM^{\mathbf{w}}(\mathbf{a}) - w_{i_0} \cdot (1 - w_{i_0}) \cdot OWA^{\mathbf{n}}(\mathbf{a}) \text{ with } \mathbf{n} = \left(0, \frac{1}{\binom{n}{2}}, \ldots, \frac{n-2}{\binom{n}{2}}, \frac{n-1}{\binom{n}{2}}\right)$$
at every $\mathbf{a} \in \mathbb{R}^n_+$ with $a_{i_0} \cdot a_{j_0} = 0$.

To prove the last two properties we benefit from the expression of the Choquet integral when the capacity is 2-additive:

$$C^{\mu}(\mathbf{a}) = \sum_{i=1}^{n} m^{\mu}(\{i\}) \cdot a_i + \sum_{\{i,j\} \subseteq X, i \neq j} m^{\mu}(\{i, j\}) \cdot \min\{a_i, a_j\} \tag{6}$$

for each $\mathbf{a} = (a_1, \ldots, a_n)$. This expression is a particular case of the general formula for the Choquet integral derived from the Möbius inverse of the capacity defining it [5, Sect. 2.6].

Now in the case where $\mathbf{w} = \left(\frac{1}{n}, \ldots, \frac{1}{n}\right)$, and $v_{i_0} = 1 - v_{j_0}$ for some $i_0 \neq j_0$, the expression in Definition 7 simplifies because

$$P^{\mathbf{w},\mathbf{v}} = \frac{\sum_{i \neq j}(w_i w_j - v_i v_j)}{\binom{n}{2}} = \frac{\frac{1}{n^2}\binom{n}{2} - v_{i_0} v_{j_0}}{\binom{n}{2}} = \frac{1}{n^2} - \frac{v_{i_0} v_{j_0}}{\binom{n}{2}}$$

therefore

$$m^{\mathbf{w},\mathbf{v}}(X') = \begin{cases} 0, & \text{when } X' = \varnothing \text{ or } X' \subseteq X, |X'| > 2 \\ w_i, & \text{when } X' = \{i\} \text{ for some } i \in X, \\ \frac{(1-v_{j_0})v_{j_0}}{\binom{n}{2}} - (1 - v_{j_0})v_{j_0}, & \text{when } X' = \{i_0, j_0\}, \\ \frac{(1-v_{j_0})v_{j_0}}{\binom{n}{2}}, & \text{otherwise.} \end{cases}$$

To prove this expression for $m^{\mathbf{w},\mathbf{v}}$, note that:

(1) If $X' = \{i_0, j_0\}$, by construction one has

$$m^{\mathbf{w},\mathbf{v}}(X') = \frac{1}{n}\frac{1}{n} - v_{i_0}v_{j_0} - P^{\mathbf{w},\mathbf{v}} = \frac{1}{n}\frac{1}{n} - v_{i_0}v_{j_0} - \left(\frac{1}{n^2} - \frac{v_{i_0}v_{j_0}}{\binom{n}{2}}\right).$$

(2) For all pairs $X' = \{i, j\}$ with $\{i, j\} \neq \{i_0, j_0\}$, one has

$$m^{\mathbf{w},\mathbf{v}}(X') = \frac{1}{n}\frac{1}{n} - 0 - P^{\mathbf{w},\mathbf{v}} = \frac{v_{i_0}v_{j_0}}{\binom{n}{2}} = \frac{(1-v_{j_0})v_{j_0}}{\binom{n}{2}}$$

For notational convenience, in the rest of the argument we denote $M = \frac{v_{i_0}v_{j_0}}{\binom{n}{2}} = \frac{(1-v_{j_0})v_{j_0}}{\binom{n}{2}}$.

When either $a_{i_0} = 0$ or $a_{j_0} = 0$, it is necessarily true that $a_{[n]} = 0$. Also, Eq. (6) produces

$$C^{\mu}(\mathbf{a}) = \sum_{i=1}^{n} m^{\mu}(\{i\}) \cdot a_i + \sum_{\substack{\{i,j\}\neq\{i_0,j_0\} \\ i\neq j}} m^{\mu}(\{i, j\}) \cdot \min\{a_i, a_j\} =$$

$$= \sum_{i=1}^{n} m^{\mu}(\{i\}) \cdot a_i + M \sum_{\substack{\{i,j\}\neq\{i_0,j_0\} \\ i\neq j}} \min\{a_i, a_j\}$$

Noting that

$$\sum_{\substack{\{i,j\}\neq\{i_0,j_0\} \\ i\neq j}} \min\{a_i, a_j\} = \sum_{i\neq j} \min\{a_i, a_j\} =$$

$$(n-1)a_{[n]} + (n-2)a_{[n-1]} + \ldots + a_{[2]} = (n-2)a_{[n-1]} + \ldots + a_{[2]}$$

we arrive at the desired conclusion.

The case $\mathbf{v} = \left(\frac{1}{n}, \ldots, \frac{1}{n}\right)$, and $w_{i_0} = 1 - w_{j_0}$ for some $i_0 \neq j_0$ is similar to the situation proven above.

4 Sufficient Conditions for the Existence of a Joint Choquet Index

In this section we resort to Chateauneuf and Jaffray's characterization (cf. Theorem 1) to demonstrate that if we impose some conditions on **w** and **v**, then it is the case that Definition 8 produces a well defined index. Examples will prove that these conditions are easy to check.

In particular, we identify two groups of sufficient conditions whose utilization is demonstrated by examples. We do this in Sects. 4.1 and 4.2 (for the first set of sufficient conditions), and then Sects. 4.3 and 4.4 (for the second set of sufficient conditions).

Our examples also prove that neither of the sufficient conditions that we provide are necessary.

4.1 First Set of Sufficient Conditions

We proceed to prove that $m^{\mathbf{w},\mathbf{v}}$ is the Möbius inverse of a capacity $\mu^{\mathbf{w},\mathbf{v}}$ under the following set of (at most $n+1$) sufficient conditions. The fact that the conditions are easy to check is shown in Sect. 4.2.

Proposition 1 *Suppose that*
(i) $\sum_{i\neq j} w_i w_j \leqslant \sum_{i\neq j} v_i v_j$ *(or equivalently,* $P^{\mathbf{w},\mathbf{v}} \leqslant 0$*), and*
(ii) when $j \in X$ *is such that* $v_j > w_j$*, the inequality* $v_j - w_j - w_j \cdot \min_{j\neq i} w_i \leqslant v_j^2$ *holds true.*
Then $m^{\mathbf{w},\mathbf{v}}$ *is the Möbius inverse of a capacity* $\mu^{\mathbf{w},\mathbf{v}}$.

Proof We can prove the claim if we check that the conditions in Theorem 1 are met.

Note that the definition of $m^{\mathbf{w},\mathbf{v}}$ guarantees that condition (1) in Theorem 1 holds true.

To prove condition (2) in Theorem 1, let us fix $X' \subseteq X$ and $j \in X'$. We need to check that the inequality $\sum_{i\in X'} m^{\mathbf{w},\mathbf{v}}(\{i,j\}) \geqslant 0$ holds true. Direct calculations show that the inequality

$$\sum_{i\in X'} m^{\mathbf{w},\mathbf{v}}(\{i,j\}) = m^{\mathbf{w},\mathbf{v}}(\{j\}) + \sum_{j\neq i\in X'} m^{\mathbf{w},\mathbf{v}}(\{i,j\}) \geqslant 0$$

is equivalent to $w_j + \sum_{j\neq i\in X'}(w_i w_j - v_i v_j) \geqslant P^{\mathbf{w},\mathbf{v}}(|X'|-1)$, or

$$w_j(1+\sum_{j\neq i\in X'} w_i) - v_j \sum_{j\neq i\in X'} v_i \geqslant P^{\mathbf{w},\mathbf{v}}(|X'|-1).$$

Suppose first $w_j - v_j \geqslant 0$. Then $w_j(1+\sum_{j\neq i\in X'} w_i) - v_j(\sum_{j\neq i\in X'} v_i) \geqslant w_j - v_j \geqslant 0 \geqslant P^{\mathbf{w},\mathbf{v}}(|X'|-1)$.

Now suppose $w_j - v_j < 0$. Using $\sum_{j\neq i\in X'} v_i \leqslant \sum_{j\neq i\in X} v_i = 1 - v_j$, we have

$$w_j\Big(1+\sum_{j\neq i\in X'} w_i\Big) - v_j\Big(\sum_{j\neq i\in X'} v_i\Big) \geqslant w_j\Big(1+\sum_{j\neq i\in X'} w_i\Big) - v_j(1-v_j) \geqslant$$

$$w_j + w_j \cdot \min_{i\neq j} w_i - v_j + v_j^2$$

and we can use assumption (ii) to conclude because

$$w_j + w_j \cdot \min_{i \neq j} w_i - v_j + v_j^2 \geqslant 0 \geqslant P^{\mathbf{w},\mathbf{v}}(|X'| - 1).$$

□

In the next section we show that neither condition (i) nor the set of conditions (ii) in Proposition 1 are necessary.

4.2 Numerical Examples

Important
In order to describe examples efficiently, as is standard, in this work the arrangements of numbers in the presentation of both capacities and their Möbius inverses replicate the Hasse diagram of the inclusion relation defined on the parts of X.

Figures 1 and 2 illustrate the diagrams that our representations replicate, in the cases $n = 3$ (together with an arbitrary capacity) and $n = 4$.

To interpret Fig. 1, the presentation of the capacity means $\mu(\{1\}) = 0.4$, $\mu(\{2\}) = 0.35$, $\mu(\{3\}) = 0.25$, $\mu(\{1, 2\}) = 0.7$, $\mu(\{1, 3\}) = 0.65$, and $\mu(\{2, 3\}) = 0.55$.

Our next example illustrates the application of the set of sufficient conditions provided in Proposition 1 in a case with $n = 4$.

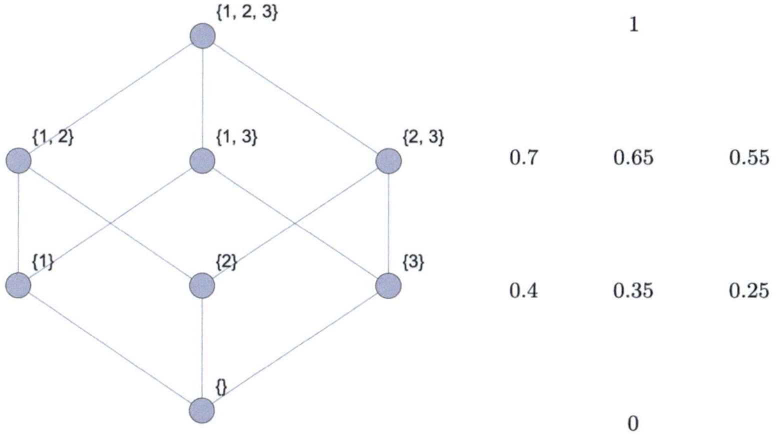

Fig. 1 Lattice of all subsets of a set with 3 elements and a presentation of an arbitrary capacity. Note the alphanumerical distribution of the subsets that carries forward to the representation

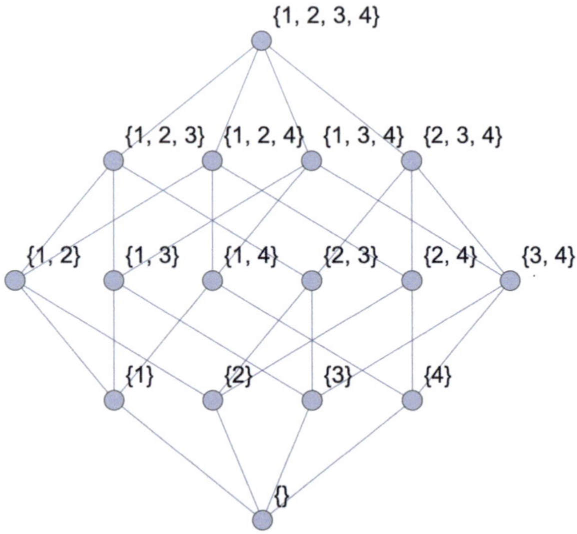

Fig. 2 Lattice of all subsets of a set with 4 elements. Note the alphanumerical distribution of the subsets

Example 1 Let us fix the weighting vectors $\mathbf{w} = (0.4, 0.2, 0.15, 0.25)$ and $\mathbf{v} = (0.3, 0.3, 0.2, 0.2)$.

Then $\sum_{i \neq j} w_i w_j = 0.3575 < 0.37 = \sum_{i \neq j} v_i v_j$, which accounts for (i) in Proposition 1.

Also, to check (ii) in Proposition 1, note that $w_1 > v_1$ and $w_4 > v_4$. Therefore we only need to do simple computations for the remaining indices, namely, $i = 2$ and $i = 3$. And indeed, $v_2^2 \geqslant v_2 - w_2 - w_2 \cdot \min_{j \neq 2} w_j$ because $0.3^2 \geqslant 0.3 - 0.2 - 0.2 \cdot 0.15$, or $0.09 \geqslant 0.07$, and $v_3^2 \geqslant v_3 - w_3 - w_3 \cdot \min_{j \neq 3} w_j$ because $0.2^2 \geqslant 0.2 - 0.15 - 0.15 \cdot 0.2$, or $0.04 \geqslant 0.2$.

Therefore the next standard presentation defines $m^{\mathbf{w},\mathbf{v}}$, which must be the Möbius inverse of a capacity (we use four-digit approximations):

$$\begin{array}{c}
0 \\
0 \quad 0 \qquad 0 \quad 0 \\
-0.0067 \ -0.0167 \ 0.0633 \quad -0.0367 \ 0.0033 \ -0.0067 \\
0.4 \quad 0.2 \qquad 0.15 \quad 0.25 \\
0
\end{array}$$

Now the Zeta transform produces the 2-additive capacity $\mu^{\mathbf{w},\mathbf{v}}$ from this expression. It can be represented as follows:

$$\begin{array}{c}
1 \\
0.64 \quad 0.96 \qquad 0.84 \quad 0.56 \\
0.5933 \ 0.4833 \ 0.7633 \quad 0.2633 \ 0.5033 \ 0.3933 \\
0.4 \quad 0.2 \qquad 0.15 \quad 0.25 \\
0
\end{array}$$

The next counterexamples prove that condition neither (i) nor (ii) in Proposition 1 are necessary.

Example 2 With $n = 3$, suppose $\mathbf{w} = (0.4, 0.35, 0.25)$ and $\mathbf{v} = (0.5, 0.3, 0.2)$. Then $\sum_{i \neq j} w_i w_j = 0.3275 > 0.31 = \sum_{i \neq j} v_i v_j$, hence (i) in Proposition 1 is contradicted.

Nevertheless we can calculate $P^{\mathbf{w},\mathbf{v}} = \frac{0.0175}{3} \approx 0.00583 > 0$. And then we can routinely define $m^{\mathbf{w},\mathbf{v}}$ as follows:

$$\begin{array}{ccc} & 0 & \\ -0.01583 & -0.00583 & 0.02167 \\ 0.4 & 0.35 & 0.25 \\ & 0 & \end{array}$$

Now we can check that the Zeta transform produces a 2-additive capacity $\mu^{\mathbf{w},\mathbf{v}}$, namely:

$$\begin{array}{ccc} & 1 & \\ 0.734167 & 0.644167 & 0.621667 \\ 0.4 & 0.35 & 0.25 \\ & 0 & \end{array}$$

Example 3 When $\mathbf{w}$ is such that $w_i = 1$ for some i, and $\mathbf{v} = (\frac{1}{n}, \ldots, \frac{1}{n})$, we explained that the joint Choquet index associated with these vectors is the projection on the i-th coordinate. Importantly, in this case (ii) in Proposition 1 is contradicted when $n > 2$, thus proving that the set of requirements (ii) is not necessary.

Note that then $\mathbf{w} = \mathbf{v}$, the conditions of Proposition 1 are met because $P^{\mathbf{w},\mathbf{v}} = 0$, and condition (ii) holds vacuously. We mentioned that the joint Choquet index associated with $\mathbf{w} = \mathbf{v}$ coincides with $WAM^{\mathbf{w}} = WAM^{\mathbf{v}}$.

4.3 Second Set of Sufficient Conditions

We now give a different condition for $m^{\mathbf{w},\mathbf{v}}$ to become the Möbius inverse of a capacity $\mu^{\mathbf{w},\mathbf{v}}$. It consists of a matrix inequality that captures several constraints simultaneously.

To pose this condition, we need further preparatory notation.

Suppose that $\mathbf{w}_{\nearrow} = (w_{\sigma(1)}, \ldots, w_{\sigma(n)})$ is a non-decreasing permutation of $\mathbf{w}$, and $\mathbf{v}_{\nearrow} = (v_{\sigma'(1)}, \ldots, v_{\sigma'(n)})$ is a non-decreasing permutation of $\mathbf{w}$. Then we define the new vectors

$$\tilde{\mathbf{w}} = (1 + w_{\sigma(1)},\ 1 + w_{\sigma(1)} + w_{\sigma(2)}, \ldots,\ 1 + w_{\sigma(1)} + w_{\sigma(2)} + \ldots + w_{\sigma(n-1)}), \quad (7)$$

$$\tilde{\mathbf{v}} = (v_{\sigma'(n)},\ v_{\sigma'(n-1)} + v_{\sigma'(n)}, \ldots,\ v_{\sigma'(2)} + v_{\sigma'(3)} + \ldots + v_{\sigma'(n)}). \quad (8)$$

We are ready to prove:

Proposition 2 *Suppose that*

$$\begin{pmatrix} \tilde{\mathbf{w}}_1 & \tilde{\mathbf{v}}_1 \\ \tilde{\mathbf{w}}_2 & \tilde{\mathbf{v}}_2 \\ \vdots & \vdots \\ \tilde{\mathbf{w}}_{n-1} & \tilde{\mathbf{v}}_{n-1} \end{pmatrix} \begin{pmatrix} \mathbf{w}_1 & \ldots & \mathbf{w}_n \\ -\mathbf{v}_1 & \ldots & -\mathbf{v}_n \end{pmatrix} \geqslant \begin{pmatrix} P^{w,v} & \ldots & P^{w,v} \\ 2P^{w,v} & \ldots & 2P^{w,v} \\ \vdots & & \vdots \\ (n-1)P^{w,v} & \ldots & (n-1)P^{w,v} \end{pmatrix}. \quad (9)$$

Then $m^{\mathbf{w},\mathbf{v}}$ is the Möbius inverse of a capacity $\mu^{\mathbf{w},\mathbf{v}}$.

Proof We need to check that the condition (2) in Theorem 1 are met, because the definition of $m^{\mathbf{w},\mathbf{v}}$ guarantees that property (1) in this characterization holds true.

To prove property (2) in Theorem 1, we need to check the inequality

$$w_j(1 + \sum_{j \neq i \in X'} w_i) - v_j \sum_{j \neq i \in X'} v_i \geqslant P^{\mathbf{w},\mathbf{v}}(|X'| - 1)$$

for any fixed $X' \subseteq X$ with at least 2 elements, and $j \in X'$ (cf., proof of Proposition 1). Denote for simplicity $k = |X'| > 1$.

Note that $\sum_{j \neq i \in X'} w_i \geqslant w_{\sigma(1)} + w_{\sigma(2)} + \ldots + w_{\sigma(k-1)}$. In words: the total of the $k-1$ numbers $\{w_i\}_{i \in X' \setminus \{j\}}$ is no less than the total of the smallest $k-1$ components of $\mathbf{w}$, i.e., the first $k-1$ components of $\mathbf{w}_{\nearrow}$.

Similarly, $\sum_{j \neq i \in X'} v_i \leqslant v_{\sigma'(n)} + w_{\sigma'(n-1)} + \ldots + w_{\sigma'(n-k+2)}$. The right-hand side of this inequality is the total of the largest $k-1$ components of $\mathbf{v}$.

Thus we can combine both inequalities to produce

$$w_j(1 + \sum_{j \neq i \in X'} w_i) - v_j \sum_{j \neq i \in X'} v_i \geqslant$$

$$w_j(1 + w_{\sigma(1)} + w_{\sigma(2)} + \ldots + w_{\sigma(k-1)}) - v_j(v_{\sigma'(n)} + w_{\sigma'(n-1)} + \ldots + w_{\sigma'(n-k+2)}) =$$

$$w_j \tilde{\mathbf{w}}_{k-1} - v_j \tilde{\mathbf{v}}_{k-1}.$$

Now we only need to couple this inequality with the condition $w_j \tilde{\mathbf{w}}_{k-1} - v_j \tilde{\mathbf{v}}_{k-1} \geqslant P^{\mathbf{w},\mathbf{v}}(|X'| - 1)$, which holds by our hypothesis as applied to row $k-1 \in \{1, 2, \ldots, n-1\}$ and column $j \in X$ of the matrix inequality. This ends the proof. □

4.4 More Numerical Examples

The next example illustrates the application of the sufficient condition in matrix form proven in Proposition 2. In passing, it demonstrates that cases not covered by Proposition 1 satisfy the requirements of this result. We work in a setting where $n = 4$.

Example 4 Suppose $\mathbf{w} = (0.15, 0.3, 0.25, 0.3)$ and $\mathbf{v} = (0.2, 0.3, 0.15, 0.35)$. Then

$$\begin{pmatrix} \mathbf{w}_1 & \dots & \mathbf{w}_n \\ -\mathbf{v}_1 & \dots & -\mathbf{v}_n \end{pmatrix} = \begin{pmatrix} 0.15 & 0.3 & 0.25 & 0.3 \\ -0.2 & -0.3 & -0.15 & -0.35 \end{pmatrix}.$$

Note that $P^{\mathbf{w},\mathbf{v}} = \frac{0.3675-0.3625}{6} \approx 0.00083$, which means that (i) in Proposition 1 is contradicted. We cannot use that result to guarantee that the joint Choquet index is well defined.

To check that the sufficient condition in Proposition 2 holds true, we compute $\mathbf{w}_{\nearrow} = (0.15, 0.25, 0.3, 0.3)$ and $\mathbf{v}_{\nearrow} = (0.15, 0.2, 0.3, 0.35)$. Now we calculate the vectors $\tilde{\mathbf{w}} = (1.15, 1.4, 1.7)$ and $\tilde{\mathbf{v}} = (0.35, 0.65, 0.85)$.

We are ready to check that Eq. (9) holds true (we use four-digit approximations):

$$\begin{pmatrix} 1.15 & 0.35 \\ 1.4 & 0.65 \\ 1.7 & 0.85 \end{pmatrix} \begin{pmatrix} 0.15 & 0.3 & 0.25 & 0.3 \\ -0.2 & -0.3 & -0.15 & -0.35 \end{pmatrix} \geqslant \begin{pmatrix} 0.0008 & 0.0008 & 0.0008 & 0.0008 \\ 0.0017 & 0.0017 & 0.0017 & 0.0017 \\ 0.0025 & 0.0025 & 0.0025 & 0.0025 \end{pmatrix}$$

which is equivalent to

$$\begin{pmatrix} 0.1025 & 0.24 & 0.235 & 0.2225 \\ 0.08 & 0.225 & 0.2525 & 0.1925 \\ 0.085 & 0.255 & 0.2975 & 0.2125 \end{pmatrix} \geqslant \begin{pmatrix} 0.0008 & 0.0008 & 0.0008 & 0.0008 \\ 0.0017 & 0.0017 & 0.0017 & 0.0017 \\ 0.0025 & 0.0025 & 0.0025 & 0.0025 \end{pmatrix}.$$

The later inequality is obviously true, and we conclude that the conditions of Proposition 2 are met.

The next standard presentation defines $m^{\mathbf{w},\mathbf{v}}$, and Proposition 2 guarantees that this set-valued function is the Möbius inverse of a capacity (we use four-digit approximations):

$$\begin{array}{cccccc} & & & 0 & & \\ & 0 & 0 & & 0 & 0 \\ -0.0158 & 0.0067 & -0.0258 & & 0.0292 & -0.0158 \quad 0.0217 \\ & 0.15 & 0.3 & & 0.25 & 0.3 \\ & & & 0 & & \end{array}$$

Finally, the 2-additive capacity $\mu^{\mathbf{w},\mathbf{v}}$ deduced from this expression can be represented as follows:

1

0.72 0.6925 0.7025 0.885

0.4342 0.4067 0.4242 0.5792 0.5842 0.5717

0.15 0.3 0.25 0.3

0

Our last example proves that the sufficient condition stated in Proposition 2 is not necessary.

Example 5 Suppose $\mathbf{w} = (0.1, 0.4, 0.2, 0.3)$ and $\mathbf{v} = (0.25, 0.2, 0.15, 0.4)$. Then

$$\begin{pmatrix} \mathbf{w}_1 & \dots & \mathbf{w}_n \\ -\mathbf{v}_1 & \dots & -\mathbf{v}_n \end{pmatrix} = \begin{pmatrix} 0.1 & 0.4 & 0.2 & 0.3 \\ -0.25 & -0.2 & -0.15 & -0.4 \end{pmatrix}.$$

Now $P^{\mathbf{w},\mathbf{v}} = -0.00125$ and we compute $\mathbf{w}_{\nearrow} = (0.1, 0.2, 0.3, 0.4)$ and $\mathbf{v}_{\nearrow} = (0.15, 0.2, 0.25, 0.4)$. Then we calculate the associated vectors $\tilde{\mathbf{w}} = (1.1, 1.3, 1.6)$ and $\tilde{\mathbf{v}} = (0.4, 0.65, 0.85)$.

We can check that Eq. (9) fails to hold true:

$$\begin{pmatrix} 1.1 & 0.4 \\ 1.3 & 0.65 \\ 1.6 & 0.85 \end{pmatrix} \begin{pmatrix} 0.1 & 0.4 & 0.2 & 0.3 \\ -0.25 & -0.2 & -0.15 & -0.4 \end{pmatrix} \ngeq \begin{pmatrix} -0.00125 & -0.00125 & -0.00125 & -0.00125 \\ -0.0025 & -0.0025 & -0.0025 & -0.0025 \\ -0.00375 & -0.00375 & -0.00375 & -0.00375 \end{pmatrix}$$

because this is equivalent to

$$\begin{pmatrix} 0.01 & 0.36 & 0.16 & 0.17 \\ -0.0325 & 0.39 & 0.1625 & 0.13 \\ -0.0525 & 0.47 & 0.1925 & 0.14 \end{pmatrix} \ngeq \begin{pmatrix} -0.00125 & -0.00125 & -0.00125 & -0.00125 \\ -0.0025 & -0.0025 & -0.0025 & -0.0025 \\ -0.00375 & -0.00375 & -0.00375 & -0.00375 \end{pmatrix}.$$

So the conditions of Proposition 2 are not met.

Nevertheless we can define $m^{\mathbf{w},\mathbf{v}}$ as follows:

0

0 0 0 0

−0.00875 −0.01625 −0.06875 0.05125 0.04125 0.00125

0.1 0.4 0.2 0.3

0

Finally, direct computations show that it defines a 2-additive capacity $\mu^{\mathbf{w},\mathbf{v}}$ represented as follows:

1

0.72625 0.76375 0.51625 0.99375

0.49125 0.28375 0.33125 0.65125 0.74125 0.50125

0.1 0.4 0.2 0.3

0

5 Computations and Comparisons

Section 4 has explored the situations for which the joint Choquet index associated with two vectores can be defined. This section computes the joint Choquet index in an example described in that section, namely, Example 1. In addition, we compare this value with the values obtained from existing indices defined from the same pair of vectors of weights.

Let us recall that Example 1 has established that for $\mathbf{w} = (0.4, 0.2, 0.15, 0.25)$ and $\mathbf{v} = (0.3, 0.3, 0.2, 0.2)$, the 2-additive capacity $\mu^{\mathbf{w},\mathbf{v}}$ is well defined. And it can be represented as follows:

		1			
	0.64	0.96	0.84	0.56	
0.5933	0.4833	0.7633	0.2633	0.5033	0.3933
	0.4	0.2	0.15	0.25	
		0			

To compute the value of the joint Choquet index that has been defined from $\mathbf{w}$ and $\mathbf{v}$ for the vector $\mathbf{a} = (14, 9, 12, 2)$, then we need to calculate $C^{\mu^{\mathbf{w},\mathbf{v}}}(14, 9, 12, 2)$ as follows:

$$C^{\mu^{\mathbf{w},\mathbf{v}}}(14, 9, 12, 2) = 8.7299 =$$

$$2\mu^{\mathbf{w},\mathbf{v}}(X) + (9-2)\mu^{\mathbf{w},\mathbf{v}}(\{1, 2, 3\}) + (12-9)\mu^{\mathbf{w},\mathbf{v}}(\{1, 3\}) + (14-12)\mu^{\mathbf{w},\mathbf{v}}(\{1\}).$$

For comparison, note that

$$WAM^{\mathbf{w}}(\mathbf{a}) = 9.7,$$

$$O^{\mathbf{v}}(\mathbf{a}) = 10,$$

$$HWA^{\mathbf{v}}(\mathbf{a}) = O^{\mathbf{v}}(4\mathbf{wa}) = O^{\mathbf{v}}(4(5.6, 1.8, 1.8, 0.5)) =$$

$$O^{\mathbf{v}}(22.4, 7.2, 7.2, 2) = 0.3 \cdot 22.4 + 0.3 \cdot 7.2 + 0.2 \cdot 7.2 + 0.2 \cdot 2 = 10.72,$$

and the OWAWA operator combines the results obtained from the $WAM^{\mathbf{w}}$ and $O^{\mathbf{v}}$ with an $\alpha \in [0, 1]$ operators, so that

$$OWAWA(\mathbf{a}) = \alpha WAM^{\mathbf{w}}(\mathbf{a}) + (1-\alpha)O^{\mathbf{v}}(\mathbf{a}) = \alpha \cdot 9.7 + (1-\alpha) \cdot 10 \in [9.7, 10].$$

In particular, in order to compute the value given by the SDOWA operator, we first need to calculate the standard deviations of both **w** and **v**, which are respectively $\sqrt{0.00875} \approx 0.0935$ and $\sqrt{0.0025} \approx 0.05$. Now we let

$$\bar{\alpha} = \frac{sd(\mathbf{w})}{sd(\mathbf{w}) + sd(\mathbf{v})} = \frac{0.0935}{0.0935 + 0.05} = 0.651568$$

and finally the evaluation is

$$\bar{\alpha} \cdot 9.7 + (1 - \bar{\alpha}) \cdot 10 = 9.80453.$$

To calculate the value of the JWA operator defined from **w** and **v** on **a**, we need first to calculate

$$\mathbf{a}_{\searrow} = (14, 12, 9, 2)$$

therefore

$$\mathbf{w}_{[\mathbf{a}]} = (0.4, 0.15, 0.2, 0.25).$$

We are ready to find

$$JWA(\mathbf{a}) = (\mathbf{w}_{[\mathbf{a}]} \oplus \mathbf{v})\mathbf{a}_{\searrow} \tag{10}$$

where we define $S = 0.4 \cdot 0.3 + 0.15 \cdot 0.3 + 0.2 \cdot 0.2 + 0.25 \cdot 0.2 = 0.255$ in order to compute

$$\mathbf{w}_{[\mathbf{a}]} \oplus \mathbf{v} = \left(\frac{0.4 \cdot 0.3}{S}, \frac{0.15 \cdot 0.3}{S}, \frac{0.2 \cdot 0.2}{S}, \frac{0.25 \cdot 0.2}{S}\right) =$$

$$(0.470588, 0.176471, 0.156863, 0.196078).$$

Therefore

$$JWA(\mathbf{a}) = (0.470588, 0.176471, 0.156863, 0.196078)(14, 12, 9, 2) = 10.5098.$$

6 Conclusions

We have explored the properties of a way to combine two vectors of weights with the help of the Choquet integral. There are similarities of the joint Choquet index that has been defined, with both weighted average means and (to a lesser degree) ordered weighted average operators that depend on n. Although joint Choquet indices generalize weighted average means, we cannot extend OWAs in general: the average mean is the only OWA operator produced by a joint Choquet index defined as in this work.

A setback of this methodology is that we cannot apply it to arbitrary pairs of vectors. Further research might disclose the exact conditions under which the joint Choquet index is well defined.

Comparisons have shown how the new index can produce evaluations below the weighted average defined from the first vector. This is in contrast with the prescriptions by other rules. Therefore, the interpretations of the roles of the two vectors of indices are also worth investigating to fully grasp the capabilities of the new index.

The computation of power indices can be studied too, inclusive of the new Υ-values defined by Torra [25].

Competing Interests The author has no conflicts of interest to declare that are relevant to the content of this chapter.

Acknowledgements The author is grateful to the Junta de Castilla y León and the European Regional Development Fund (Grant CLU-2O25-2-03) for the financial support to the Research Unit of Excellence "Economic Management for Sustainability" (GECOS). Thanks are given to the participants of the XXVI edition of the international congress Mathematical Modelling in Engineering & Human Behaviour (MME&HB2024) for useful comments.

References

1. Alcantud, J.C.R., de Andrés, R.: The problem of collective identity in a fuzzy environment. Fuzzy Sets Syst. **315**, 57–75 (2017)
2. Alcantud, J.C.R., Santos-García, G., Akram, M.: OWA aggregation operators and multi-agent decisions with N-soft sets. Expert Syst. Appl. **203**, 117430 (2022)
3. Alcantud, J.C.R., Santos-García, G., Akram, M.: A novel methodology for multi-agent decision-making based on N-soft sets. Soft. Comput. (2023). https://doi.org/10.1007/s00500-023-08522-0
4. Beliakov, G.: Comparing apples and oranges: The weighted OWA function. Int. J. Intell. Syst. **33**(5), 1089–1108 (2017)
5. Beliakov, G., Pradera, A., Calvo, T.: Aggregation Functions: A Guide for Practitioners. Springer, Heidelberg (2007)
6. Broomell, S.B., Wagner, C.: The Joint Weighted Average (JWA) Operator. arXiv:2302.11885v1 [cs.AI]. Cited 23 Feb 2023
7. Cardin, M., Giove, S.: SDOWA: A new OWA operator for decision making. In: Esposito, A., et al. (eds.) Progresses in Artificial Intelligence and Neural Systems, pp. 305–315. Springer Singapore, Singapore (2021)
8. Chateauneuf, A., Jaffray, J.-Y.: Some characterizations of lower probabilities and other monotone capacities through the use of Möbius inversion. Math. Soc. Sci. **17**(3), 263–283 (1989)
9. Choquet, G.: Theory of capacities. Ann. I. Fourier **5**, 131–295 (1953)
10. Gajdos, T.: Measuring inequalities without linearity in envy: Choquet integrals for symmetric capacities. J. Econ. Theory **106**(1), 190–200 (2002)
11. Grabisch, M.: k-order additive discrete fuzzy measures and their representation. Fuzzy Sets Syst. **92**(2), 167–189 (1997)
12. Grabisch, M.: Set Functions, Games and Capacities in Decision Making. Springer Nature (2016)

13. Grabisch, M., Labreuche, C.: A decade of application of the Choquet and Sugeno integrals in multi-criteria decision aid. Ann. Oper. Res. **175**(1), 247–290 (2010)
14. Llamazares, B.: An analysis of some functions that generalizes weighted means and OWA operators. Int. J. Intell. Syst. **28**(4), 380–393 (2013)
15. Llamazares, B.: Constructing Choquet integral-based operators that generalize weighted means and OWA operators. Inf. Fusion **23**, 131–138 (2015)
16. Llamazares, B.: SUOWA operators: Constructing semi-uninorms and analyzing specific cases. Fuzzy Sets Syst. **287**, 119–136 (2016)
17. Llamazares, B.: SUOWA operators: an analysis of their conjunctive/disjunctive character. Fuzzy Sets Syst. **357**, 117–134 (2019)
18. Llamazares, B.: Generalizations of weighted means and OWA operators by using unimodal weighting vectors. IEEE Trans. Fuzzy Syst. **28**(9), 1961–1970 (2020)
19. Merigò, J.M.: A unified model between the weighted average and the induced OWA operator. Expert Syst. Appl. **38**, 11560–11572 (2011)
20. Mesiar, R., Mesiarová, R.: Fuzzy integrals–What are they? Int. J. Intell. Syst. **23**(2), 199–212 (2008)
21. Miranda, P., García-Segador, P.: Combinatorial structure of the polytope of 2-additive measures. IEEE Trans. Fuzzy Syst. **28**(11), 2864–2874 (2020)
22. Pelegrina, G.D., Duarte, L.T., Grabisch, M.: A k-additive Choquet integral-based approach to approximate the SHAP values for local interpretability in machine learning. Artif. Intell. **325**, 104014 (2023)
23. Sugeno, M.: Theory of fuzzy integrals and applications. Ph.D. thesis, Tokyo Institute of Technology (1974)
24. Torra, V.: The weighted OWA operator. Int. J. Intell. Syst. **12**, 153–166 (1997)
25. Torra, V.: Υ-values: power indices à la orness for non-additive measures. IEEE Trans. Fuzzy Syst. **32**, 4099–4108 (2024)
26. Torra, V., Godo, L.: Continuous WOWA operators with application to defuzzification. In: Calvo, T., Mayor, G., Mesiar, R. (eds.) Aggregation Operators. Studies in Fuzziness and Soft Computing, vol 97, pp- 159–176. Physica-Verlag, Heidelberg (2002)
27. Torra, V., Lv, Z.: On the WOWA operator and its interpolation function. Int. J. Intell. Syst. **24**, 1039–1056 (2009)
28. Torra, V., Narukawa, Y.: The h-index and the number of citations: two fuzzy integrals. IEEE Trans. Fuzzy Syst. **16**(3), 795–797 (2008)
29. Xu, Z.S., Da, Q.L.: An overview of operators for aggregating information. Int. J. Intell. Syst. **18**, 953–969 (2003)
30. Yager, R.R.: On ordered weighted averaging aggregation operators in multicriteria decision making. IEEE Trans. Syst. Man Cybern. **18**, 183–190 (1988)
31. Yager, R.R., Alajlan, N.: A generalized framework for mean aggregation: toward the modeling of cognitive aspects. Inf. Fusion **17**, 65–73 (2014)

Combined Matrices of Sign Regular Matrices. The Case 3-by-3

Begoña Cantó, Rafael Cantó, and Ana M. Urbano

Abstract This work explores sign regular (SR) matrices, a key class in matrix theory with applications in mathematics, statistics, engineering, economics, and computer design. A SR matrix is defined as a matrix whose all minors of the same order with identical signs, with specific subclasses determined by the matrix signature. Among these, totally positive (TP) and totally negative (TN) matrices are particularly important. We investigate the properties and subclasses of SR matrices, focusing on their connection to combined matrices, checkerboard matrices, and doubly quasi-stochastic matrices. A central objective is to determine if there exists a nonsingular matrix in specific SR subclasses whose combined matrix is both doubly quasi-stochastic and strictly checkerboard or strictly inverse checkerboard. The chapter introduces key concepts and presents new results, including a characterization of these matrices based on their signatures. Theoretical results are grounded in existing literature and extended through novel constructions. Further chapters provide detailed exploration of specific SR subclasses and their properties, ensuring a comprehensive analysis of these structured matrices.

Keywords Sign regular matrices · Totally positive and totally negative matrices · Combined matrices · Doubly quasi-stochastic matrices

1 Introduction and Background

In this chapter we study sign regular (SR) matrices, an important class of matrices in matrix theory due to its applications. A real matrix is SR if all its minors of the same

B. Cantó (✉) · R. Cantó · A. M. Urbano
I.U. de Matemática Multidisciplinar, Universitat Politècnica de València, Camí de Vera s/n, València, Spain
e-mail: bcanto@mat.upv.es

R. Cantó
e-mail: rcanto@mat.upv.es

A. M. Urbano
e-mail: amurbano@mat.upv.es

J. R. Torregrosa (ed.), *Mathematical Modelling in Engineering Human Behaviour*, Springer Proceedings in Mathematics & Statistics 513,
https://doi.org/10.1007/978-3-032-00567-0_7

order have the same sign. These matrices are useful in many areas of mathematics, statistics, engineering, economics, computer desing, etc. (see [1, 2]). Some authors have been studied this topics, for example, in [3] the authors develop accurate and efficient solutions for linear systems involving generalized sign regular matrices with a particular signature. In [4] a more efficient or simplified method for testing sign regular matrices compared to more traditional methods is provided. In [5], the authors introduce the specific aspects such as characterization, perturbation analysis, and interval properties of sign regular matrices and clarify what each term means in the context of them and why understanding these properties is important. Finally, in [6] new theoretical insights or characterizations regarding the combination of almost strictly sign regular matrices are defined and how these matrices affects their sign regularity and under what conditions certain properties are preserved are introduced.

The signature of an SR matrix reflects a highly structured behavior of the matrix. Recall the following definition.

Definition 1 ([7]) Given a *signature* ϵ, that is, a real sequence $\epsilon = (\epsilon_1, \ldots, \epsilon_k, \ldots, \epsilon_n)$ such that $\mid \epsilon_k \mid = 1$ for all $1 \leq k \leq n$, we say that an $n \times n$ real matrix A is *sign regular* with signature ϵ, if for all $1 \leq k \leq n$ and $\alpha_1, \alpha_2 \in Q_{k,n}$, we have

$$\epsilon_k \det A(\alpha_1, \alpha_2) > 0$$

where $Q_{k,n}$ denotes the set of all increasing sequences of k natural numbers less than or equal to n and $A(\alpha_1, \alpha_2)$ is the submatrix of A containing rows numbered by α_1 and columns numbered by α_2.

The signature of matrix A shows valuable information about the sign of its determinants with ϵ_i, $i = 1, 2 \ldots, n$, which can each take on values of 1 or -1.

Remark 1 If we consider sign regular matrices of order 3, then we have 8 possible subclasses:

$$\begin{array}{llll}(1,\ 1,\ 1), & (1,\ 1,\ -1), & (1,\ -1,\ 1), & (1,\ -1,\ -1),\\ (-1,\ 1,\ 1), & (-1,\ 1,\ -1), & (-1,\ -1,\ 1), & (-1,\ -1,\ -1).\end{array}$$

Among these SR matrices, two particularly important classes are (1, 1, 1) that corresponds to the totally positive (TP) matrices of order 3, with all their minors of any order positive and $(-1, -1, -1)$ that corresponds to the totally negative (TN) matrices of order 3, that is, matrices with all their minors of any order negative. The accurate solutions of different numerical linear algebra problems concerning TP and TN matrices have been studied by several authors For example, the approach to solving linear systems with structured matrices by means of the bidiagonal factorization of TP matrices is considered in [8], the use of a differential equation as a tool for the study of TP matrices is shown. In [9], necessary and sufficient conditions for three real numbers greater than 1 to be the entries on the main diagonal of the combined matrix of a totally positive matrix A are given, and a complex method for obtaining A is proposed in [10], some conditions that the entries of a specific doubly quasi-stochastic matrix of order 3 must satisfy in order to exist a real matrix such that its combined matrix is that one are given, and an easy technique to obtain a totally positive matrix from the study of the variation interval of one of its entries is provided.

Historically (see the note of Fallat [11]), the investigations into oscillatory and mechanical systems of Gantmacher and Krein introduced an important class of structured matrices called oscillatory matrices. An $n \times n$ matrix A is called *totally nonnegative* if every minor of A is nonnegative, see [12], and an $n \times n$ totally nonnegative matrix is called *oscillatory* if some positive integral power of it is TP. They proved that every oscillatory matrix have distinct positive eigenvalues and they studied some properties on eigenvectors, including the number of sign changes among the components of each eigenvector.

Matrices with all their minors negative are called TN and they have recently studied by several authors. In [13] they studied characterizations of these classes of matrices by minors, by their full rank factorization and by their thin QR factorization. An $n \times m$ matrix A is called *totally nonpositive* if all its minors are nonpositive and TN if all its minors are negative. For square matrices, spectral properties and the LDU factorization are analyzed in [14], and a characterization in terms of the parameters of the Neville elimination is obtained in [15]. The totally nonpositive and the TN matrices can be considered as an extension of the partially nonpositive matrices, p.n.p. (partially negative matrices, p.n.), that is, matrices such that all its principals minors are nonpositive (negative). A p.n. matrix is called N-matrix in economic models [16]. Also, the p.n. matrices arise in conjunction with Lemke's algorithm for solving linear and convex quadratic programming problems.

Now, we recall the next definition related with the 8 subclasses of matrices given in Remark 1.

Definition 2 The *combined matrix* of a real nonsingular matrix A is the matrix $C(A) = A \circ A^{-t}$, where $\circ$ means the Hadamard (entrywise) product, and A^{-t} means the inverse transpose, $(A^t)^{-1}$, of A.

In control theory, Bristol [17] introduced the Relative Gain Array (RGA) as the combined matrix of a real matrix A, and it is used to determine the best pairing that corresponds to good controller performance. That is, the RGA is a normalized form of the gain matrix that describes the impact of each control variable on the output relative to each control variable's impact on other variables. The process interaction of open-loop and closed-loop control systems is measured for all possible input-output variable pairings. A ratio of this open-loop 'gain' to this closed-loop 'gain' is determined, and the results are displayed in the RGA matrix. Concretely, the RGA is a classical widely-used method for determining the best input-output pairing for multivariable process control systems. It has been considered in many practical open-loop and closed-loop control applications and in the analysis of many fundamental steady-state closed-loop system properties such as stability and robustness. Concretely, in some chemical engineering applications a diagonal of RGA in which the entries are near 1 is used to determine the pairing of inputs and outputs for further design analysis (see, for instance, [18, 19]), where many applications in control theory of the RGA matrices are for small sizes of order $n = 2, 3$ or 4 (see [20, 21]).

The following Lemma gives basic properties of these matrices.

Lemma 1 ([22]) *The combined matrix* $C(A) = (c_{ij}) = A \circ A^{-T}$ *of a nonsingular matrix* $A = (a_{ij})$ *satisfies*

(*a*) *If* D_1 *and* D_2 *are two diagonal nonsingular matrices then* $C(A) = C(D_1 A D_2)$.
(*b*) *If* P *and* Q *are two permutation matrices, then* $C(PAQ) = PC(A)Q$.
(*c*) $C(A^{-T}) = C(A)$.
(*d*) $C(-A) = C(A)$.
(*e*) *If* $c_{ij} \neq 0$ *then* $a_{ij} \neq 0$, $\forall\, i, j$.

Some properties of combined matrices have been studied in Mathematics. For instance, the problem of characterizing when three real numbers are the diagonal entries of a combined matrix can be found in [23–26]. In [27], we study doubly stochastic combined matrices and we characterize the real matrices of order 3 such that their combined matrix is a given doubly stochastic matrix. In this paper, doubly quasi-stochastic matrices appear as the natural generalization of doubly stochastic matrices, and we have characterized conditions that the entries of a specific doubly quasi-stochastic matrix of order 3 must satisfy, for there to exist a real matrix such that its combined matrix is that one, considering two cases: when all entries of the doubly quasi-stochastic matrix are known or when only the main diagonal entries of it are known.

By Remark 1 and Lemma 1, the 8 possible subclasses of sign regular matrices of order 3 can be grouped into 4 subclasses: (a) $\epsilon_1 = (1,\ 1,\ 1), \epsilon_2 = (-1,\ 1,\ -1)$; ($b$) $\epsilon_3 = (1,\ -1,\ -1), \epsilon_4 = (-1,\ -1,\ 1)$; ($c$) $\epsilon_5 = (1,\ -1,\ 1), \epsilon_6 = (-1,\ -1,\ -1)$; ($d$) $\epsilon_7 = (1,\ 1,\ -1), \epsilon_8 = (-1,\ 1,\ 1)$. This is because, by property (d) of Lemma 1, the combined matrix of the matrices A and $-A$ is the same and the signature of both matrices are opposite with the exception of the second entry because the minors of order 2 do not change.

Note that, the combined matrices from subclasses (a) and (b) are doubly quasi-stochastic and strictly checkerboard matrices, while those from subclasses (c) and (d) are doubly quasi-stochastic and strictly inverse checkerboard matrices. We recall these definitions.

Definition 3 A real matrix $U = (u_{ij})$ of order n is a *checkerboard matrix* if $\text{sign}(u_{ij}) = (-1)^{i+j}$ or $u_{ij} = 0\, \forall\, i, j = 1, \ldots, n$. U is a *strictly checkerboard matrix* when no entry is zero [23].

Additionally, U is an *inverse checkerboard matrix* if $\text{sign}(u_{ij}) = (-1)^{i+j+1}$ or $u_{ij} = 0\, \forall\, i, j = 1, \ldots, n$. U is a *strictly inverse checkerboard matrix* when no entry is zero.

Definition 4 ([10]) A real matrix $U = (u_{ij})$ of order n is a *doubly quasi-stochastic matrix* if the values in each row and each column sum up to 1. If the matrix is nonnegative matrix, it is called a *doubly stochastic matrix*.

Using a similar technique as in [10], we want to achieve the objective of this work. It is to determine if there exists a nonsingular matrix A whose signature is in the subclasses (c) or (d), such that its combined matrix is a doubly quasi-stochastic and checkerboard matrix and to construct it. To ensure the work is self-contained, the next section will present the most relevant results for the matrices in the subclasses (a) and (b). Section 3 will address the matrices in the subclass (c) and (d). Finally, in Sect. 4, we summarize de obtained results.

2 Doubly Quasi-stochastic and Strictly Checkerboard Matrices

In this section we consider the combined matrices from the subclasses (a) whose signature are $\epsilon_1 = (1,\ 1,\ 1), \epsilon_2 = (-1,\ 1,\ -1)$ and (b) with $\epsilon_3 = (1,\ -1,\ -1), \epsilon_4 = (-1,\ -1,\ 1)$. In these cases, their combined matrices are doubly quasi-stochastic and strictly checkerboard matrices, and we consider the following question.

Question 1. What are the conditions for a doubly quasi-stochastic and strictly checkerboard matrix U to be the combined matrix of a nonsingular matrix A that belongs to the subclasses (a) or (b)?

To answer the question we use the following result from [10].

Theorem 1 *Let $U = (u_{ij})$ be a doubly quasi-stochastic matrix of order 3 without zero entries in its first row and its first column. Then, there exists a nonsingular matrix $A = (a_{ij})$, such that $C(A) = U$, if and only if the polynomial*

$$P(x) = u_{11}x^2 + (u_{22} - u_{13}u_{31} - u_{33}u_{11})x + u_{33}u_{13}u_{31} \tag{1}$$

has at least a real root. Additionally, if $u_{21} + u_{31} \neq 0$, the real root is different from $\dfrac{-u_{13}u_{31}}{u_{21} + u_{31}}$.

If $u_{ii} > 0$, for $i = 1, 2, 3$, $u_{11} - u_{22} + u_{33} - 1 = -K < 0$ and $u_{13} \in\ ]0, K[$, then U is a doubly quasi-stochastic and checkerboard matrix,

$$U = \begin{pmatrix} u_{11} & 1 - u_{11} - u_{13} & u_{13} \\ 1 - u_{11} + u_{13} - K & u_{22} & 1 - u_{13} - u_{33} \\ K - u_{13} & 1 - u_{33} + u_{13} - K & u_{33} \end{pmatrix}.$$

Moreover, if $u_{13} \notin \mathcal{J}$ where $\mathcal{J} = \{0, K, 1 - u_{11}, K + u_{11} - 1, 1 - u_{33}, K + u_{33} - 1\}$, then U is strictly checkerboard.

Example 1 Consider the doubly quasi-stochastic matrix U,

$$U = \begin{pmatrix} 0.5 & 0.5 - u_{13} & u_{13} \\ u_{13} - 14.8 & 15 & 0.8 - u_{13} \\ 15.3 - u_{13} & u_{13} - 14.5 & 0.2 \end{pmatrix}.$$

Since $u_{11} - u_{22} + u_{33} - 1 = -15.3 < 0$, for $u_{13} \in]0, 15.3[$ and $u_{13} \notin \{0.5, 14.8, 0.8, 14.5\}$ the matrix U is strictly checkerboard. For example, for $u_{13} = 12$ we have the doubly quasi-stochastic and strictly checkerboard matrix

$$U = \begin{pmatrix} 0.5 & -11.5 & 12 \\ -2.8 & 15 & -11.2 \\ 3.3 & -2.5 & 0.2 \end{pmatrix}.$$

Since U is a doubly quasi-stochastic and strictly checkerboard matrix by Lemma 1, if $A = (a_{ij})$ satisfies that $C(A) = U$, then $a_{ij} \neq 0$, $\forall\, i, j$, and it can be factorized as $A = D_1^{-1} T D_2^{-1}$ where D_1 and D_2 are nonsingular diagonal matrices and T and $S = T^{-1}$ are given by

$$T = \begin{pmatrix} 1 & 1 & 1 \\ 1 & t_{22} & t_{23} \\ 1 & t_{32} & t_{33} \end{pmatrix}, \quad S = \begin{pmatrix} u_{11} & u_{21} & u_{31} \\ u_{12} & u_{13} - u_{21} + s_{33} & -u_{31} - s_{33} \\ u_{13} & -u_{13} - s_{33} & s_{33} \end{pmatrix}, \tag{2}$$

where s_{33} is a root of the polynomial given in (1).

The answer to Question 1 for the subclass (a) is given in the following result.

Theorem 2 ([10]) *Necessary and sufficient conditions for a doubly quasi-stochastic and strictly checkerboard matrix $U = (u_{ij})$ of order 3 to be the combined matrix of a nonsingular matrix A with signature $\epsilon_1 = (1,\ 1,\ 1)$ are:*

1. *The entries u_{11}, u_{22} and u_{33} satisfy*

$$u_{ii} > 1, \quad i = 1, 2, 3$$
$$u_{11} + u_{33} - 1 < u_{22} < u_{11}u_{33}.$$

2. *The entry $u_{13} \in]\,0,\ K\,[$, $K = -(u_{11} - u_{22} + u_{33} - 1) = u_{13} + u_{31}$, satisfies*

$$u_{13}(K - u_{13}) \leq (\sqrt{u_{22}} - \sqrt{u_{11}u_{33}})^2.$$

Remark 2 Since $C(-A) = C(A) = U$ Theorem 2 also holds for $-A$ with signature ϵ_2.

Example 2 Consider the doubly quasi-stochastic and checkerboard matrix

$$U = \begin{pmatrix} 8 & -7 - u_{13} & u_{13} \\ -40 + u_{13} & 120 & -79 - u_{13} \\ 33 - u_{13} & -112 + u_{13} & 80 \end{pmatrix}, \ \forall u_{13} \in]0,\ 33[.$$

Note that, $u_{ii} > 1,\ i = 1, 2, 3,\ u_{11} + u_{33} - 1 = 87 < u_{22} = 120 < u_{11}u_{33} = 640$, $u_{11} - u_{22} + u_{33} - 1 = -33 = -K$ and $(\sqrt{u_{22}} - \sqrt{u_{11}u_{22}})^2 = 205.7437$.

Choosing $u_{13} = 30$, we have $u_{13}(12 - u_{13}) = 90 < 205.7437$ and the doubly quasi-stochastic and strictly checkerboard matrix

$$U = \begin{pmatrix} 8 & -37 & 30 \\ -10 & 120 & -109 \\ 3 & -82 & 80 \end{pmatrix}.$$

The polynomial $P(x)$ given in (1) is $P(x) = 8x^2 - 610x + 7200$ and it has the positive roots $x_1 = 61.6519$ and $x_2 = 14.5981$. As $u_{21} + u_{31} = -1 \neq 0$ we can construct the matrix S using x_1 or x_2.

- For $x_1 = 61.6519$, we have the matrices

$$S_1 = \begin{pmatrix} 8 & -10 & 3 \\ -37 & 101.6519 & -64.65194 \\ 30 & -91.6519 & 61.6519 \end{pmatrix}, \quad T_1 = \begin{pmatrix} 1 & 1 & 1 \\ 1 & 1.1805 & 1.1893 \\ 1 & 1.2683 & 1.2976 \end{pmatrix},$$

with $\det(S_1) = 341.5634,\ \det(T_1) = 0.0029$ and $\epsilon_{T_1} = \epsilon_1 = (1,\ 1,\ 1)$.

- For $x_2 = 14.5981$, we obtain the matrices

$$S_2 = \begin{pmatrix} 8 & -10 & 3 \\ -37 & 54.5981 & -17.5981 \\ 30 & -44.5981 & 14.5981 \end{pmatrix}, \quad T_2 = \begin{pmatrix} 1 & 1 & 1 \\ 1 & 2.1979 & 2.4441 \\ 1 & 4.6596 & 5.4802 \end{pmatrix},$$

with $\det(S_2) = 12.1866$, $\det(T_2) = 0.0821$ and $\epsilon_{T_2} = \epsilon_1 = (1,\ 1,\ 1)$.

Note that,

$$\tilde{T}_1 = -T_1 = \begin{pmatrix} -1 & -1 & -1 \\ -1 & -1.1805 & -1.1893 \\ -1 & -1.2683 & -1.2976 \end{pmatrix}, \ \det(\tilde{T}_1) = -0.0029,$$

$$\tilde{T}_2 = -T_2 = \begin{pmatrix} -1 & -1 & -1 \\ -1 & -2.1979 & -2.4441 \\ -1 & -4.6596 & -5.4802 \end{pmatrix}, \ \det(\tilde{T}_2) = -0.0821,$$

and, in these cases, $\epsilon_{\tilde{T}_1} = \epsilon_{\tilde{T}_2} = \epsilon_2 = (-1,\ 1,\ -1)$.

From T_1 and T_2 we can obtain matrices A whose signature is $\epsilon_A = \epsilon_1 = (1,\ 1,\ 1)$ that is, TP matrices, and $C(A) = U$, using diagonal matrices D_1 and D_2 with all diagonal entries positive. For example, if $D_1 = \text{diag}(2, 1, 3)$ and $D_2 = \text{diag}(3, 2, 5)$ we have the matrix

$$A = D_1\, T_1\, D_2 = \begin{pmatrix} 6 & 4 & 10 \\ 3 & 2.361 & 5.9464 \\ 9 & 7.610 & 19.4641 \end{pmatrix},$$

whose signature is $\epsilon_A = \epsilon_1 = (1,\ 1,\ 1)$. The same result is obtained using matrices D_1 and D_2 with all diagonal entries negative. However, if D_1 (D_2) has all diagonal entries positive and D_2 (D_1) has all diagonal entries negative we obtain matrices A whose combined matrix is U and their signature is $\epsilon_A = \epsilon_2 = (-1,\ 1,\ -1)$. For example, if $D_1 = \text{diag}(2, 1, 3)$ and $D_2 = \text{diag}(-3, -2, -5)$ we obtain the matrix

$$A = D_1\, T_1\, D_2 = \begin{pmatrix} -6 & -4 & -10 \\ -3 & -2.361 & -5.9464 \\ -9 & -7.610 & -19.4641 \end{pmatrix},$$

whose signature is $\epsilon_A = \epsilon_2 = (-1,\ 1,\ -1)$.

Analogously, from $\tilde{T}_1$ and $\tilde{T}_2$ we obtain matrices A whose combined matrix is U and with signature $\epsilon_A = \epsilon_2 = (-1,\ 1,\ -1)$ or $\epsilon_A = \epsilon_1 = (1,\ 1,\ 1)$ using diagonal matrices D_1 and D_2 with all diagonal entries positive or negative or with one of them with diagonal entries positive and the other one with diagonal entries negative.

Now, we consider the subclass (b). By Lemma 1, $C(T) = C(A) = U$ and using a similar technique as in [10], we give the following result, which it has already been presented in a Mathematics Spanish Conference, but a sketch of the proof is included here.

Theorem 3 *Necessary and sufficient conditions for a doubly quasi-stochastic and strictly checkerboard matrix $U = (u_{ij})$ of order 3 to be the combined matrix of a nonsingular matrix A with signature $\epsilon_3 = (1,\ -1,\ -1)$ are:*

1. *The entries u_{11}, u_{22} and u_{33} satisfy*

$$\begin{aligned} &u_{ii} \geq 1, \quad i = 1, 2, 3 \\ &u_{22} > max\{1, u_{11}, u_{33}, u_{11} + u_{33} - 1\}. \end{aligned} \tag{3}$$

2. *The entry $u_{13} \in]0,\ K\,[,\ K = -(u_{11} - u_{22} + u_{33} - 1) = u_{13} + u_{31}$, satisfies*

$$u_{13}(K - u_{13}) \geq (\sqrt{u_{22}} + \sqrt{u_{11}u_{33}})^2. \tag{4}$$

Proof Let $U = (u_{ij})$ be a doubly quasi-stochastic and strictly checkerboard matrix of order 3 where its entries satisfy the conditions given in (3) and (4). Applying Theorem 3 of [10], there exists a matrix A such that $C(A) = U$. We rewrite the polynomial (1) in terms of K as

$$P(x) = u_{11}x^2 + (u_{22} - u_{13}(K - u_{13}) - u_{11}u_{33})x + u_{33}u_{13}(K - u_{13})$$

and obtain its real positive roots. Thus, the matrices S and T given in (2) are a checkerboard matrix and a positive matrices, respectively. To determine if $\epsilon_T = \epsilon_3 = (1,\ -1,\ -1)$, we study the sign of $\det(S)$, and to do so, we study the sign of $\det(S[1, 3])$. As in [10], we obtain $\det(S[1, 3]) < 0$. Thus, $\det(S) < 0$ and therefore $\epsilon_T = \epsilon_3 = (1,\ -1,\ -1)$.

From T and using diagonal matrices D_1 and D_2, we obtain a diagonal equivalent totally positive matrix A such that $C(A) = U$, with $\epsilon_3 = (1,\ -1,\ -1)$.

Conversely, let T be a sign regular matrix of order 3 with $\epsilon_T = \epsilon_3 = (1,\ -1,\ -1)$ given in (2) such that $0 < t_{ij} < 1,\ i, j = 2, 3,\ t_{33} < \min\{t_{23}, t_{32}\}$ and $t_{22} > \max\{t_{23}, t_{32}\}$. Thus, $0 < t_{33} < t_{22} < 1$. Let $U = (u_{ij})$ be the combined matrix of T. Considering that $\epsilon_T = (1, -1, -1)$, then U is a doubly quasi-stochastic and checkerboard matrix. Applying [10], we have $u_{11} + u_{33} - 1 < u_{22}$.

Since $\det(T) = t_{22}(t_{33} - 1) + t_{23}(1 - t_{32}) + (t_{32} - t_{33}) < 0$ then $u_{22} = \dfrac{t_{22}(t_{33} - 1)}{\det(T)} > 1$. Moreover $u_{11} < u_{22}$ and $u_{33} < u_{22}$. Thus, $u_{22} > \max\{1, u_{11}, u_{22}, u_{11} + u_{33} - 1\}$.

Additionally, we prove that $u_{13}(K - u_{13}) \geq (\sqrt{u_{22}} + \sqrt{u_{11}u_{33}})^2$ using reduction ad absurdum. For that, we suppose that

$$u_{13}(K - u_{13}) < (\sqrt{u_{22}} + \sqrt{u_{11}u_{33}})^2 = u_{22} + 2\sqrt{u_{11}u_{22}u_{33}} + u_{11}u_{33}$$

holds, and in this case we obtain that $\epsilon_T \neq \epsilon_3 = (1,\ -1,\ -1)$.

Thus, $u_{13}(K - u_{13}) \geq (\sqrt{u_{22}} + \sqrt{u_{11}u_{33}})^2$. □

Remark 3 Since $C(-A) = C(A) = U$ Theorem 3 also holds for $-A$ with signature ϵ_4.

Example 3 Consider the doubly quasi-stochastic and checkerboard matrix

$$U = \begin{pmatrix} 2 & -1 - u_{13} & u_{13} \\ -13 + u_{13} & 15 & -1 - u_{13} \\ 12 - u_{13} & -13 + u_{13} & 2 \end{pmatrix},\quad \forall u_{13} \in]0,\ 12[.$$

Note that, $u_{11} + u_{33} - 1 = 3$, $u_{22} = 15 > \max\{1, u_{11}, u_{11} + u_{33} - 1\} = 3$ and $(\sqrt{u_{22}} + \sqrt{u_{11}u_{22}})^2 = 34.4919$.

Choosing $u_{13} = 6$, we have $u_{13}(12 - u_{13}) = 36 > 34.4919$ and the following doubly quasi-stochastic and strictly checkerboard matrix

$$U = \begin{pmatrix} 2 & -7 & 6 \\ -7 & 15 & -7 \\ 6 & -7 & 2 \end{pmatrix}.$$

The polynomial $P(x)$ given in (1) is $P(x) = 2x^2 - 25x + 72$ and it has the positive roots $x_1 = 8$ and $x_2 = 9/2$. As $u_{21} + u_{31} = -1 \neq 0$ we can construct the matrix S using x_1 or x_2.

- For $x_1 = 8$, we have the matrices

$$S_1 = \begin{pmatrix} 2 & -7 & 6 \\ -7 & 21 & -14 \\ 6 & -14 & 8 \end{pmatrix}, \quad T_1 = \begin{pmatrix} 1 & 1 & 1 \\ 1 & 0.7143 & 0.5 \\ 1 & 0.5 & 0.25 \end{pmatrix},$$

with $\det(S_1) = -28$, $\det(T_1) = -1/28$ and $\epsilon_{T_1} = \epsilon_3 = (1, \ -1, \ -1)$.

- For $x_2 = 9/2$, we have the matrices

$$S_2 = \begin{pmatrix} 2 & -7 & 6 \\ -7 & 17.5 & -10.5 \\ 6 & -10.5 & 4.5 \end{pmatrix}, \quad T_2 = \begin{pmatrix} 1 & 1 & 1 \\ 1 & 0.8571 & 0.6667 \\ 1 & 0.6667 & 0.4444 \end{pmatrix},$$

with $\det(S_2) = -31.5$, $\det(T_2) = -0.0317$ and $\epsilon_{T_2} = \epsilon_3 = (1, \ -1, \ -1)$.

Note that,

$$\tilde{T}_1 = -T_1 = \begin{pmatrix} -1 & -1 & -1 \\ -1 & -0.7143 & -0.5 \\ -1 & -0.5 & -0.25 \end{pmatrix}, \quad \det(\tilde{T}_1) = 0.036,$$

$$\tilde{T}_2 = -T_2 = \begin{pmatrix} -1 & -1 & -1 \\ -1 & -0.8571 & -0.6667 \\ -1 & -0.6667 & -0.4444 \end{pmatrix}, \quad \det(\tilde{T}_2) = 0.0317,$$

and, in these cases, $\epsilon_{\tilde{T}_1} = \epsilon_{\tilde{T}_2} = \epsilon_4 = (-1, \ -1, \ 1)$.

From T_1 and T_2 we can obtain matrices A whose signature is $\epsilon_A = \epsilon_3 = (1, \ -1, \ -1)$ and $C(A) = U$, using diagonal matrices D_1 and D_2 with all diagonal entries positive. For example, if $D_1 = \text{diag}(2, 1, 3)$ and $D_2 = \text{diag}(3, 2, 5)$ we have the matrix

$$A = D_1 \, T_1 \, D_2 = \begin{pmatrix} 6 & 4 & 10 \\ 3 & 1.43 & 2.5 \\ 9 & 3 & 3.75 \end{pmatrix},$$

whose signature is $\epsilon_A = \epsilon_3 = (1, \ -1, \ -1)$. The same result is obtained using matrices D_1 and D_2 with all diagonal entries negative. If D_1 (D_2) has all diagonal entries positive and D_2 (D_1) has all diagonal entries negative we obtain matrices A whose combined matrix is U and its signature is $\epsilon_A = \epsilon_4 = (-1,\ -1,\ 1)$.

Analogously, from $\tilde{T}_1$ and $\tilde{T}_2$ we obtain matrices A whose combined matrix is U and their signature is $\epsilon_A = \epsilon_4 = (-1,\ -1,\ 1)$ or $\epsilon_A = \epsilon_3 = (1,\ -1,\ -1)$ using diagonal matrices D_1 and D_2 with all diagonal entries positive or negative or with one of them with diagonal entries positive and the other with diagonal entries negative.

3 Doubly Quasi-stochastic and Strictly Inverse Checkerboard Matrices

In this section we consider the combined matrices from the subclasses (c) whose signature are $\epsilon_5 = (1,\ -1,\ 1)$, $\epsilon_6 = (-1,\ -1,\ -1)$ and (d) with $\epsilon_7 = (1,\ 1,\ -1)$, $\epsilon_8 = (-1,\ 1,\ 1)$. In these cases their combined matrices are doubly quasi-stochastic and strictly inverse checkerboard matrices, and we consider the following question.

Question 2. What are the conditions for a doubly quasi-stochastic and strictly inverse checkerboard matrix U to be the combined matrix of a nonsingular matrix A that belongs to the subclasses (c) *and* (d)*?*

If $U = (u_{ij})$ is doubly quasi-stochastic and strictly inverse checkerboard then $sign(u_{ij}) = (-1)^{i+j+1}$ and

$$u_{11} - u_{22} + u_{33} - 1 = -(u_{13} + u_{31}) > 0.$$

Note that, in this case, $u_{11} < 0, u_{33} < 0$ and therefore $u_{22} < -1$. Hence, we shown the following result to answer Question 2.

Theorem 4 *A matrix $U = (u_{ij})$ of order 3 is doubly quasi-stochastic and strictly inverse checkerboard if and only if it is the combined matrix of a nonsingular matrix A with signature $\epsilon_5 = (1,\ -1,\ 1)$ or $\epsilon_7 = (1,\ 1,\ -1)$.*

Proof Let $U = (u_{ij})$ be a doubly quasi-stochastic and strictly inverse checkerboard matrix of order 3. Then, it is verify that $u_{11} - u_{22} + u_{33} - 1 = -(u_{13} + u_{31}) > 0$ with u_{11}, u_{33}, u_{13} and u_{31} less that zero and $u_{22} < -1$.

Since $u_{11}u_{22}u_{33} < 0$ applying [10, Theorem 6], there exists a matrix A such that $C(A) = U$. As $K = u_{13} + u_{31}$, the polynomial given in (1) is

$$P(x) = u_{11}x^2 + (u_{22} - u_{13}(K - u_{13}) - u_{11}u_{33})x + u_{33}u_{13}(K - u_{13})$$

and its real roots are

$$x = \frac{-(u_{22} - u_{13}(K - u_{13}) - u_{11}u_{33})}{2u_{11}}$$
$$\pm \frac{\sqrt{(u_{22} - u_{13}(K - u_{13}) - u_{11}u_{33})^2 - 4u_{11}u_{33}u_{13}(K - u_{13})}}{2u_{11}}.$$

Since $u_{22} - u_{13}(K - u_{13}) - u_{11}u_{33} < 0$ and $u_{22} < u_{11}u_{33}$ the numerator is always positive. Therefore, the roots x_1 and x_2 are negative. If we factorized matrix A using (2), we obtain that the matrices S_1 and S_2 constructed from x_1 and x_2 are strictly inverse checkerboard, and consequently the matrices $T_1 = S_1^{-1}$ and $T_2 = S_2^{-1}$ are positive. To study the signature of T_1 and T_2 we determine the sign of $\det(S_1[1, 3])$ and $\det(S_2[1, 3])$.

Let x_1 be the real root of $P(x)$ obtained by adding a partial sum of the square root. Then

$$\det(S_1[1,3]) = u_{11}x_1 - u_{13}(K - u_{13}) = \frac{-u_{22} - u_{13}(K - u_{13}) + u_{11}u_{33}}{2} +$$
$$+ \frac{\sqrt{(u_{22} - u_{13}(K - u_{13}) - u_{11}u_{33})^2 - 4u_{11}u_{33}u_{13}(K - u_{13})}}{2}.$$

If $-u_{22} - u_{13}(K - u_{13}) + u_{11}u_{33} > 0$ then $\det(S_1[1, 3]) > 0$ and $\det(S_1) > 0$. Thus, $\epsilon_{T_1} = (1,\ -1,\ 1)$. Therefore $\epsilon_A = \epsilon_5 = (1,\ -1,\ 1)$.

If $-u_{22} - u_{13}(K - u_{13}) + u_{11}u_{33} < 0$ we study the sign. We denote

$$B = -(-u_{22} - u_{13}(K - u_{13}) + u_{11}u_{33}) = u_{22} + u_{13}(K - u_{13}) - u_{11}u_{33}$$
$$C = \sqrt{(u_{22} - u_{13}(K - u_{13}) - u_{11}u_{33})^2 - 4u_{11}u_{33}u_{13}(K - u_{13})}$$

with $B > 0$ and $C > 0$. To ensure that the determinant is negative, we prove $B < C$, or equivalently, $B^2 < C^2$. In fact,

$$B^2 - C^2 = 4u_{22}u_{13}(K - u_{13}) < 0 \rightarrow B < C \rightarrow 0 < -B + C.$$

Then $\det(S_1[1, 3]) > 0$ and $\det(S_1) > 0$. Thus, $\epsilon_{T_1} = (1,\ -1,\ 1)$. Therefore $\epsilon_A = \epsilon_5 = (1,\ -1,\ 1)$.

Let x_2 be the real root of $P(x)$ obtained by subtracting a partial sum of the square root. Then

$$\det(S_2[1,3]) = u_{11}x_2 - u_{13}(K - u_{13}) = \frac{-u_{22} - u_{13}(K - u_{13}) + u_{11}u_{33}}{2} +$$
$$- \frac{\sqrt{(u_{22} - u_{13}(K - u_{13}) - u_{11}u_{33})^2 - 4u_{11}u_{33}u_{13}(K - u_{13})}}{2}.$$

If $-u_{22} - u_{13}(K - u_{13}) + u_{11}u_{33} > 0$ and

$$B = -u_{22} - u_{13}(K - u_{13}) + u_{11}u_{33}$$
$$C = \sqrt{(u_{22} - u_{13}(K - u_{13}) - u_{11}u_{33})^2 - 4u_{11}u_{33}u_{13}(K - u_{13})}$$

it is straightforward that $B^2 < C^2 \rightarrow B < C \rightarrow B - C < 0$. Then $\det(S_2[1, 3]) < 0$, and $\det(S_2) < 0$. Thus, $\epsilon_{T_2} = (1, 1, -1)$. Therefore $\epsilon_A = \epsilon_7 = (1, 1, -1)$.

If $-u_{22} - u_{13}(K - u_{13}) + u_{11}u_{33} < 0$ then $\det(S_2[1, 3]) < 0$. Reasoning as the previous case we prove that $\epsilon_{T_2} = (1, 1, -1)$. Therefore $\epsilon_A = \epsilon_7 = (1, 1, -1)$.

Conversely, let A be a sign regular matrix of order 3 with $\epsilon_A = \epsilon_7 = (1, 1, -1)$. Using (2) we have $0 < t_{ij} < 1, i, j = 2, 3, t_{33} < \min\{t_{23}, t_{32}\}$ and $t_{22} > \max\{t_{23}, t_{32}\}$. Thus, $0 < t_{33} < t_{22} < 1$. Let $U = (u_{ij})$ be the combined matrix of T. Considering that $\epsilon_T = (1, -1, -1)$, then U is a doubly quasi-stochastic and strictly inverse checkerboard matrix. Applying [10], we have $u_{11} + u_{33} - 1 < u_{22}$.

Since $\det(T) = t_{22}(t_{33} - 1) + t_{23}(1 - t_{32}) + (t_{32} - t_{33}) < 0$ then $t_{22}(t_{33} - 1) < \det(T) < 0$ and $u_{22} = \dfrac{t_{22}(t_{33} - 1)}{\det(T)} > 1$. Moreover

$$u_{11} = \frac{t_{22}t_{33} - t_{23}t_{32}}{\det(T)} < \frac{t_{22}(t_{33} - 1)}{\det(T)} = u_{22}$$
$$u_{33} = \frac{t_{33}(t_{22} - 1)}{\det(T)} < \frac{t_{22}(t_{33} - 1)}{\det(T)} = u_{22}.$$

Thus, $u_{22} > \max\{1, u_{11}, u_{22}, u_{11} + u_{33} - 1\}$.

Now, we prove that $u_{13}(K - u_{13}) \geq (\sqrt{u_{22}} + \sqrt{u_{11}u_{33}})^2$ using reduction ad absurdum. We suppose that

$$u_{13}(K - u_{13}) < (\sqrt{u_{22}} + \sqrt{u_{11}u_{33}})^2 = u_{22} + 2\sqrt{u_{11}u_{22}u_{33}} + u_{11}u_{33}$$

holds. Then,

$$2\sqrt{u_{11}u_{22}u_{33}} > u_{13}(K - u_{13}) - u_{11}u_{33} - u_{22}.$$

If $|u_{13}(K - u_{13}) - u_{11}u_{33} - u_{22}| < 2\sqrt{u_{11}u_{22}u_{33}}$ working in the same way that in [10], we obtain that $P(x)$ given in (1) has no real roots. As a consequence, does not exist any nonsingular matrix A such that $C(A) = U$.

If $|u_{13}(K - u_{13}) - u_{11}u_{33} - u_{22}| > 2\sqrt{u_{11}u_{22}u_{33}}$ then

$$-u_{13}(K - u_{13}) + u_{11}u_{33} + u_{22} > 2\sqrt{u_{11}u_{22}u_{33}},$$

which implies that

$$u_{13}(K - u_{13}) < u_{11}u_{33} + u_{22} - 2\sqrt{u_{11}u_{22}u_{33}} = (\sqrt{u_{22}} - \sqrt{u_{11}u_{33}})^2.$$

In this case, and analogously to [10], we obtain $\det(S) > 0$. As a consequence, $\epsilon_T \neq \epsilon_3 = (1, -1, -1)$.

Thus, $u_{13}(K - u_{13}) \geq (\sqrt{u_{22}} + \sqrt{u_{11}u_{33}})^2$ and therefore U is a doubly quasi-stochastic and strictly inverse checkerboard matrix such that $C(A) = U$. □

Remark 4 Since $C(-A) = C(A) = U$ Theorem 4 also holds for $-A$ with signature ϵ_6 or ϵ_8.

Example 4 Consider the following doubly quasi-stochastic and inverse checkerboard matrix

$$U = \begin{pmatrix} -3 & 5 & -1 \\ 6 & -10 & 5 \\ -2 & 6 & -3 \end{pmatrix}.$$

In this case we have

$$-K = u_{11} - u_{22} + u_{33} - 1 = 3 > 0 \text{ and } -u_{22} - u_{13}(K - u_{13}) + u_{11}u_{33} = 10 - 2 + 9 = 17 > 0.$$

The second degree polynomial associated with U, $P(x) = -3x^2 - 21x - 6$ has the negative roots:

$$x_1 = \frac{21 + \sqrt{(-21)^2 - 4(-3)(-6)}}{-6} \approx -6.7016$$
$$x_2 = \frac{21 - \sqrt{(-21)^2 - 4(-3)(-6)}}{-6} \approx -0.2984.$$

The inverse checkerboard matrices S_i and the corresponding matrices T_i are:

- If $x_1 = -6.7016$, then

$$S_1 = \begin{pmatrix} -3 & 6 & -2 \\ 5 & -13.7016 & 8.7016 \\ -1 & 7.7016 & -6.7016 \end{pmatrix} \text{ and } T_1 = \begin{pmatrix} 1 & 1 & 1 \\ 1 & 0.7298 & 0.6492 \\ 1 & 0.6895 & 0.4477 \end{pmatrix}$$

with $\det(S_1) = 24.8062$, $\det(T_1) = 0.0403$ and $\epsilon_{T_1} = \epsilon_5 = (1, -1, 1)$.

- If $x_2 = -0.2984$, then

$$S_2 = \begin{pmatrix} -3 & 6 & -2 \\ 5 & -7.2984 & 2.2984 \\ -1 & 1.2984 & -0.2984 \end{pmatrix} \text{ and } T_2 = \begin{pmatrix} 1 & 1 & 1 \\ 1 & 1.3702 & 3.8508 \\ 1 & 2.6105 & 10.0523 \end{pmatrix}$$

with $\det(S_2) = -0.8062$, $\det(T_2) = -1.2403$ and $\epsilon_{T_2} = \epsilon_7 = (1, 1, -1)$.

Note that, the matrix

$$\tilde{T}_1 = -T_1 = \begin{pmatrix} -1 & -1 & -1 \\ -1 & -0.7298 & -0.6492 \\ -1 & -0.6895 & -0.4477 \end{pmatrix}$$

has $\epsilon_{\tilde{T}_1} = \epsilon_6 = (-1,\ -1,\ -1)$. Therefore, $\tilde{T}_2$ is a TN matrix.

The matrix

$$\tilde{T}_2 = -T_2 = \begin{pmatrix} -1 & -1 & -1 \\ -1 & -1.3702 & -3.8508 \\ -1 & -2.6105 & -10.0523 \end{pmatrix}$$

has $\epsilon_{\tilde{T}_1} = \epsilon_8 = (-1,\ 1,\ 1)$.

From T_1 we can obtain matrices A whose signature is $\epsilon_A = \epsilon_5 = (1,\ -1,\ 1)$ and $C(A) = U$, using diagonal matrices D_1 and D_2 with all diagonal entries positive or negative. However, if D_1 (D_2) has all diagonal entries positive and D_2 (D_1) has all diagonal entries negative we obtain matrices A whose combined matrix is U and their signature is $\epsilon_A = \epsilon_6 = (-1,\ -1,\ -1)$. For example, if $D_1 = \mathrm{diag}(2, 1, 3)$ and $D_2 = \mathrm{diag}(-3, -2, -5)$ we have the matrix

$$A = D_1\, T_1\, D_2 = \begin{pmatrix} -6 & -4 & -10 \\ -3 & -1.4597 & -3.2461 \\ -9 & -4.1372 & -6.7149 \end{pmatrix}$$

whose signature is $\epsilon_A = \epsilon_6 = (-1,\ -1,\ -1)$. Note that A is a TN matrix.

From $\tilde{T}_1$ we obtain matrices A whose combined matrix is U and with signature $\epsilon_A = \epsilon_6 = (-1,\ -1,\ -1)$, that is TN matrices, or $\epsilon_A = \epsilon_5 = (1,\ -1,\ 1)$ using diagonal matrices D_1 and D_2 with all diagonal entries positive or negative or with one of them with diagonal entries positive and the other with diagonal entries negative.

From T_2 we can obtain matrices A whose signature is $\epsilon_A = \epsilon_7 = (1,\ 1,\ -1)$ and $C(A) = U$, using diagonal matrices D_1 and D_2 with all diagonal entries positive or negative. If D_1 (D_2) has all diagonal entries positive and D_2 (D_1) has all diagonal entries negative we obtain matrices A whose combined matrix is U and its signature is $\epsilon_A = \epsilon_8 = (-1,\ 1,\ 1)$.

Finally, from $\tilde{T}_2$ we obtain matrices A whose combined matrix is U and their signature is $\epsilon_A = \epsilon_8 = (-1,\ 1,\ 1)$ or $\epsilon_A = \epsilon_7 = (1,\ 1,\ -1)$ using diagonal matrices D_1 and D_2 with all diagonal entries positive or negative or with one of them with diagonal entries positive and the other with diagonal entries negative.

Example 5 Now, we consider the following doubly quasi-stochastic and inverse checkerboard matrix

$$U = \begin{pmatrix} -2 & 11 & -8 \\ 7 & -20 & 14 \\ -4 & 10 & -5 \end{pmatrix}.$$

In this case we have

$$-K = u_{11} - u_{22} + u_{33} - 1 = 12 \text{ and } -u_{22} - u_{13}(K - u_{13}) + u_{11}u_{33} = 20 - 32 + 10 = -2 < 0.$$

The second degree polynomial associated with U, $P(x) = -2x^2 - 62x - 160$ has the negative roots: $x_1 \approx -28.159$ and $x_2 \approx -2.841$.

The inverse checkerboard matrices S_i and the corresponding matrices T_i are:

- If $x_1 = -28.159$, then

$$S_1 = \begin{pmatrix} -2 & 7 & -4 \\ 11 & -43.159 & 32.159 \\ -8 & 36.159 & -28.159 \end{pmatrix} \text{ and } T_1 = \begin{pmatrix} 1 & 1 & 1 \\ 1 & 0.4634 & 0.3872 \\ 1 & 0.3110 & 0.1776 \end{pmatrix}$$

with $\det(S_1) = 52.477$, $\det(T_1) = 0.0191$ and $\epsilon_{T_1} = \epsilon_5 = (1, -1, 1)$.

- If $x_2 = -2.841$, then

$$S_2 = \begin{pmatrix} -2 & 7 & -4 \\ 11 & -17.841 & 6.841 \\ -8 & 10.841 & -2.841 \end{pmatrix} \text{ and } T_2 = \begin{pmatrix} 1 & 1 & 1 \\ 1 & 1.1210 & 1.2914 \\ 1 & 1.4618 & 1.7599 \end{pmatrix}$$

with $\det(S_2) = -23.477$, $\det(T2) = -0.0426$ and $\epsilon_{T_2} = \epsilon_7 = (1, 1, -1)$.

As in Example 4, the matrix

$$\tilde{T}_1 = -T_1 = \begin{pmatrix} -1 & -1 & -1 \\ -1 & -0.4634 & -0.3872 \\ -1 & -0.3110 & -0.1776 \end{pmatrix}$$

is totally negative and therefore its signature is $\epsilon_6 = (-1, -1, -1)$ and the matrix

$$\tilde{T}_2 = -T_2 = \begin{pmatrix} -1 & -1 & -1 \\ -1 & -1.1210 & -1.2914 \\ -1 & -1.4618 & -1.7599 \end{pmatrix}$$

has signature $\epsilon_{\tilde{T}_2} = \epsilon_8 = (-1, 1, 1)$.

Similarly to Example 4, starting from T_1 we obtain matrices A whose signature is $\epsilon_A = \epsilon_5 = (1, -1, 1)$ and for which $C(A) = U$. This is achieved by employing diagonal matrices D_1 and D_2, where all diagonal entries are either positive or negative. However, if D_1 has all positive diagonal entries and D_2 entirely negative ones, or vice versa, we obtain matrices A for which the combined matrix is U and the signature becomes $\epsilon_A = \epsilon_6 = (-1, -1, -1)$. In this case, the resulting matrices are TN.

Likewise, from $\tilde{T}_1$, we can also obtain matrices A whose combined matrix is U, with a signature of either $\epsilon_A = \epsilon_6 = (-1,\ -1,\ -1)$, corresponding to TN matrices, or $\epsilon_A = \epsilon_5 = (1,\ -1,\ 1)$. This is accomplished using diagonal matrices D_1 and D_2 where both have either all diagonal positive or negative entries, or where one has positive diagonal entries and the other has negative ones.

Starting from T_2, we construct matrices A with a signature $\epsilon_A = \epsilon_7 = (1,\ 1,\ -1)$ and combined matrix $C(A) = U$, using diagonal matrices D_1 and D_2 where all diagonal entries are either positive or negative. If D_1 has all positive diagonal entries and D_2 has all negative diagonal entries (or vice versa), the resulting matrices A have a combined matrix U and a signature is $\epsilon_A = \epsilon_8 = (-1,\ 1,\ 1)$.

Finally, from $\tilde{T}_2$ we obtain matrices A whose combined matrix is U and whose signature is either $\epsilon_A = \epsilon_8 = (-1,\ 1,\ 1)$ or $\epsilon_A = \epsilon_7 = (1,\ 1,\ -1)$. This is achieved by using diagonal matrices D_1 and D_2 with either all positive or negative diagonal entries, or with one matrix having positive diagonal entries and the other negative diagonal ones.

Remark 5 In [23] the authors distinguish between the symmetric and nonsymmetric cases. In our work, we do not make this distinction because if the matrix A is symmetric with signature $\epsilon_5 = (1,\ -1,\ 1)$ or $\epsilon_7 = (1,\ 1,\ -1)$, then the matrix T will also be symmetric and maintain the same signature. As a result, U will be symmetric, doubly quasi-stochastic, and an inverse checkerboard matrix. For the inverse result, the proof of Theorem 4 does not depend on whether U is symmetric or not. The construction of matrix S ensures that it will be symmetric if U is symmetric, and consequently, T will also be symmetric.

4 Conclusions

In this work we consider 4 subclasses of sign regular matrices A: (a) $\epsilon_1 = (1,\ 1,\ 1)$, $\epsilon_2 = (-1,\ 1,\ -1)$; (b) $\epsilon_3 = (1,\ -1,\ -1)$, $\epsilon_4 = (-1,\ -1,\ 1)$; (c) $\epsilon_5 = (1,\ -1,\ 1)$, $\epsilon_6 = (-1,\ -1,\ -1)$; (d) $\epsilon_7 = (1,\ 1,\ -1)$, $\epsilon_8 = (-1,\ 1,\ 1)$. For each subclass we characterize when a doubly quasi-stochastic, strictly checkerboard or strictly inverse checkerboard matrix is a combined matrix of a nonsingular matrix A.

The following outline summarizes the main results obtained.

$U = (u_{ij})$ doubly quasi-stochastic matrix

$$\Downarrow$$

$$u_{11} - u_{22} + u_{33} - 1 = -(u_{13} + u_{31})$$

$\swarrow$ $\searrow$

checkerboard	inverse checkerboard
$\Downarrow$	$\Downarrow$
$u_{11}, u_{22}, u_{33} > 0$	$u_{11}, u_{33} < 0,\ u_{22} < -1$
$u_{11} - u_{22} + u_{33} - 1 < 0$	$u_{11} - u_{22} + u_{33} - 1 > 0$
$\Downarrow$	$\Downarrow$

If $u_{ij} \neq 0$, for all $i, j = 1, 2, \ldots, n$

$\Downarrow$	$\Downarrow$
strictly checkerboard	strictly inverse checkerboard

Since $C(-A) = C(A)$, we show the following results.

$U = (u_{ij})$ doubly quasi-stochastic and strictly checkerboard matrix

$\Downarrow$	$\Downarrow$
Theorem 2	Theorem 3
$\Downarrow$	$\Downarrow$
$U = C(A)\ /\ \epsilon_1 = (1,\ 1,\ 1)$	$U = C(A)\ /\ \epsilon_3 = (1,\ -1,\ -1)$
or	or
$U = C(B)\ /\ \epsilon_2 = (-1,\ 1,\ -1)$	$U = C(B)\ /\ \epsilon_4 = (-1,\ -1,\ 1)$
$\Updownarrow$	$\Updownarrow$
$u_{ii} > 1,\ i = 1, 2, 3$	$u_{ii} \geq 1,$
$u_{11} + u_{33} - 1 < u_{22} < u_{11}u_{33}$	$u_{22} > \max\{1,\ u_{11},\ u_{33},\ u_{11} + u_{33} - 1\}$
$u_{13}u_{31} \leq \left(\sqrt{u_{22}} - \sqrt{u_{11}u_{33}}\right)^2$	$u_{13}u_{31} \geq \left(\sqrt{u_{22}} + \sqrt{u_{11}u_{33}}\right)^2$

$$\boxed{\text{Theorem 4}}$$
$$\Downarrow$$

$$\boxed{\begin{array}{c} U = (u_{ij}) \text{ doubly quasi-stochastic and strictly inverse checkerboard} \\ \\ \Updownarrow \\ \\ \begin{array}{c||c} U = C(A) \; / \; \epsilon_5 = (1,\ -1,\ 1) & U = C(A) \; / \; \epsilon_7 = (1,\ 1,\ -1) \\ \text{or} & \text{or} \\ U = C(B) \; / \; \epsilon_6 = (-1,\ -1,\ -1) & U = C(B) \; / \; \epsilon_8 = (-1,\ 1,\ 1) \end{array} \end{array}}$$

Competing Interests The authors have no conflicts of interest to declare that are relevant to the content of this chapter.

Ethics Approval Not apply.

Acknowledgements This research was partially supported by Conselleria de Innovación, Universidades, Ciencia y Sociedad Digital de la Generalitat Valenciana under Spanish grant CIAICO/2021/162.

References

1. Huang, R.: Componentwise error analysis for linear system associated with sign regular matrices. J. Comput. Appl. Math. **255**, 133–141 (2014)
2. Alonso, P., Peña, J.M., Serrano, M.L.: On the characterization of almost sign regular matrices. J. Comput. Appl. Math. **275**, 480–488 (2015)
3. Yang, Z., Chen, T.: The accurate and efficient solutions of linear systems for generalized sign regular matrices with certain signature. J. Comput. Appl. Math. (2023). https://doi.org/10.1016/j.cam.2023.115280
4. Choudhury, P.N., Yadav, S.: Sign regular matrices and variation diminution: single-vector tests and characterizations, following Schoenberg, Gantmacher-Krein, and Motzkin. In: Proceedings in Press American Mathematical Society (2024). https://doi.org/10.1090/proc/17026
5. Adm, M., Garloff, J.: Characterization, perturbation, and interval property of certain sign regular matrices. Linear Alg. Appl. (2021). https://doi.org/10.1016/j.laa.2020.12.007
6. Alonso, P., Peña, J.M., Serrano, M.L.: Combined matrices of almost strictly sign regular matrices. J. Comput. Appl. Math. **354**, 144–151 (2019)
7. Peña, J.M.: On nonsingular sign regular matrices. Linear Alg. Appl. (2015). https://doi.org/10.1016/S0024-3795(02)00437-8
8. Martinez, J.J.: The application of the bidiagonal factorization of totally positive matrices in numerical linear algebra. Axioms (2024). https://doi.org/10.3390/axioms13040258
9. Goodman, G.S.: Analytical and Probabilistic Methods for Totally Positive Matrices. Mathematics and its Applications. Springer (2021)
10. Boix, M., Cantó, B., Cantó, R., Urbano, A.M.: The range of combined matrices and doubly quasi-stochastic matrices of order 3. Linear Algebra Appl. (2024). https://doi.org/10.1016/j.laa.2024.04.028

11. Fallat, S.M.: A remark on oscillatory matrices. Linear Alg. and Appl. (2004). https://doi.org/10.1016/j.laa.2003.12.006
12. Minc, H.: Nonnegative Matrices. John Wiley and Sons Inc. (1988)
13. Cantó, R., Ricarte, B., Urbano, A.M.: Some characterizations of totally nonpositive (totally negative) matrices. Electron. J. Linear Algebr. **20**, 241–253 (2010)
14. Fallat, S.M., Van Den Driessche, P.: On matrices with allminors negative. Electron. J. Linear Algebr. **7**, 92–99 (2000)
15. Gasca, M., Peña, J.M.: A test for strict sign-regularity. Linear Algebr. Appl. **197/198** 133–142 (1994)
16. Bapat, R.B., Raghavab, T.E.S.: Nonnegative Matrices and Applications. Cambridge University Press (1997)
17. Bristol, E.: On a new measure of interaction for multivariable process control. IEEE Trans. Autom. Control (1966). https://doi.org/10.1109/TAC.1966.1098266
18. Mousavi, M., Haeri, M.: Welding current and arc voltage control in a GMAW process using ARMarkov based MPC. Control Engineering Practice (2011). https://doi.org/10.1016/j.conengprac.2011.07.015
19. Skogestad, S., Morari, M.: Implications of large RGA elements on control performance. Ind. Eng. Chem. Res. (1987). https://doi.org/10.1021/ie00071a025
20. Chiu, M.S.: A methodology for the synthesis of robust decentralized control systems [dissertation]. Georgia Institute of Technology, Atlanta GA (1991)
21. Hovd, M., Skogestad, S.: Sequencial Desig of Descentralized Controllers. Automatica (1994). https://doi.org/10.1016/0005-1098(94)90099-X
22. Horn, R.A., Johnson, C.R.: Topics in Matrix Analysis. Cambridge University Press (1991)
23. Bru, R., Gassó, M.T., Giménez, I., Santana, M.: Diagonal entries of the combined matrix of a totally negative matrix. Linear Multilinear Alg. (2017). https://doi.org/10.1080/03081087.2016.1261079
24. Fiedler, M.: Relations between the diagonal entries of an M-matrix and its inverse. Mat. Fyz Casopis. **12**(2), 123–128 (1962)
25. Fiedler, M.: Relations between the diagonal entries of two mutually inverse positive definite matrices. Czechoslovak Math. J. (1964). https://doi.org/10.21136/CMI.1964.100600
26. Fiedler, M., Markham, T.L.: Combined matrices in special classes of matrices. Linear Algebr. Appl. (2011). https://doi.org/10.1016/j.laa.2011.03.054
27. Cantó, B., Cantó, R., Gassó, M.T., Urbano, A.M.: Doubly stochastic and combined matrices. Linear Multilinear Algebr. (2024). https://doi.org/10.1080/03081087.2023.2260073

MIX
Papier aus verantwortungsvollen Quellen
Paper from responsible sources
FSC® C105338

If you have any concerns about our products,
you can contact us on
ProductSafety@springernature.com

In case Publisher is established outside the EU,
the EU authorized representative is:
Springer Nature Customer Service Center GmbH
Europaplatz 3, 69115 Heidelberg, Germany

Printed by Libri Plureos GmbH
in Hamburg, Germany